T0214019

SpringerBriefs in Mathematics

SpringerBriefs in Mathematics showcases expositions in all areas of mathematics and applied mathematics. Manuscripts presenting new results or a single new result in a classical field, new field, or an emerging topic, applications, or bridges between new results and already published works, are encouraged. The series is intended for mathematicians and applied mathematicians. All works are peer-reviewed to meet the highest standards of scientific literature.

BCAM SpringerBriefs

BCAM *SpringerBriefs* aims to publish contributions in the following disciplines: Applied Mathematics, Finance, Statistics and Computer Science. BCAM has appointed an Editorial Board, who evaluate and review proposals.

Typical topics include: a timely report of state-of-the-art analytical techniques, bridge between new research results published in journal articles and a contextual literature review, a snapshot of a hot or emerging topic, a presentation of core concepts that students must understand in order to make independent contributions.

Please submit your proposal to the Editorial Board or to Francesca Bonadei, Executive Editor Mathematics, Statistics, and Engineering: francesca.bonadei@springer.com.

More information about this series at http://www.springer.com/series/10030

Philipp Braun · Lars Grüne · Christopher M. Kellett

(In-)Stability of Differential Inclusions

Notions, Equivalences, and Lyapunov-like Characterizations

Philipp Braun
School of Engineering
Australian National University
Canberra, Australia

Lars Grüne
Mathematical Institute
Universität Bayreuth
Bayreuth, Bayern, Germany

Christopher M. Kellett
School of Engineering
Australian National University
Canberra, Australia

ISSN 2191-8198   ISSN 2191-8201 (electronic)
SpringerBriefs in Mathematics
ISBN 978-3-030-76316-9   ISBN 978-3-030-76317-6 (eBook)
https://doi.org/10.1007/978-3-030-76317-6

This Springer imprint is published by the registered company Springer Nature Switzerland AG
The registered company address is: Gewerbestrasse 11, 6330 Cham, Switzerland

Preface

The fundamental theory that emerged from Aleksandr Mikhailovich Lyapunov's doctoral thesis [50] more than 100 years ago has been and still is the main tool to analyze stability properties of dynamical systems. Lyapunov or Lyapunov-like functions are monotone functions when evaluated along the solution of a dynamical system. Based on the monotonicity property, stability or instability of invariant sets can be concluded without the need to derive explicit solutions of the system dynamics.

In this monograph, existing results characterizing stability and stabilizability of the origin of differential inclusions through Lyapunov and control Lyapunov functions are reviewed and new characterizations for instability and destabilization characterized through Lyapunov-like arguments are derived. To distinguish between stability and instability, stability results are characterized through Lyapunov and control Lyapunov functions whereas instability is characterized through Chetaev and control Chetaev functions. In addition, similarities and differences between stability and instability and stabilizability and destabilizability of the origin of a differential inclusion are summarized. These connections are established by considering dynamics in forward time, in backward time, or by considering a scaled version of the differential inclusion. In total, the diagram shown in Fig. 1.1 is obtained, unifying new and existing results in a consistent notation.

As a last contribution of the monograph, ideas combining control Lyapunov and control Chetaev functions into a single framework are discussed. Through this approach, convergence (i.e., stability) and avoidance (i.e., instability) are guaranteed simultaneously.

The genesis of this monograph emerged from the preliminary results in [11], published as a conference paper in the proceedings of the 57th IEEE Conference on Decision and Control. Additionally, the ideas combining properties of control Lyapunov and control Chetaev functions rely on conference papers [13] and [14].

The authors would like to thank the anonymous reviewers for their helpful detailed comments and insights and for Example 2.17 and Remark 4.7 in particular.

Newcastle (Mulubinba), Australia   Philipp Braun
Bayreuth, Germany   Lars Grüne
Newcastle (Mulubinba), Australia   Christopher M. Kellett
March 2021

Contents

About the Authors

Philipp Braun received his Ph.D. in Applied Mathematics from the University of Bayreuth, Bayreuth, Germany, in 2016. He is a Research Fellow in the School of Engineering at the Australian National University, Canberra, Australia. His main research interests include control theory with a focus on stability analysis of nonlinear systems using Lyapunov methods and model predictive control.

Lars Grüne received his Ph.D. in Mathematics from the University of Augsburg, Augsburg, Germany, in 1996 and the Habilitation from Goethe University Frankfurt, Frankfurt, Germany, in 2001. He is currently a Professor of Applied Mathematics with the University of Bayreuth, Bayreuth, Germany. His research interests include mathematical systems and control theory with a focus on numerical and optimization-based methods for nonlinear systems.

Christopher M. Kellett received his Ph.D. in electrical and computer engineering from the University of California, Santa Barbara, in 2002. He is currently a Professor and the Director of the School of Engineering at the Australian National University. His research interests are in the general area of systems and control theory with an emphasis on the stability, robustness, and performance of nonlinear systems with applications in both social and technological systems.

Chapter 1
Introduction

Abstract Lyapunov methods have been and still are one of the main tools to analyze stability properties of dynamical systems. In this monograph Lyapunov results
characterizing stability and stabilizability of the origin of differential inclusions are
reviewed. To characterize instability and destabilizability, Lyapunov-like functions,
called Chetaev and control Chetaev functions in the monograph, are introduced.
Based on their definition and by mirroring existing results on stability, analogue
results for instability are derived. Moreover, by looking at the dynamics of a differential inclusion in backward time, similarities and differences between stability
of the origin in forward time and instability in backward time, and vice versa, are
discussed. Similarly, invariance of stability and instability properties of equilibria
of differential equations with respect to a scaling are summarized. As a final result,
ideas combining control Lyapunov and control Chetaev functions to simultaneously
guarantee stability, i.e., convergence, and instability, i.e., avoidance, are outlined.
The work is addressed at researchers working in control as well as graduate students
in control engineering and applied mathematics.

Keywords Lyapunov methods · Differential inclusions · Stability of nonlinear
systems · Instability of nonlinear systems · Stabilization/destabilization of
nonlinear systems · Stabilizability and destabilizability

Lyapunov functions, originating from the work by Aleksandr Mikhailovich Lyapunov
at the end of the 19th century [50], have been the main tool for the stability analysis
of dynamical systems for more than a century. The strength of Lyapunov functions
is that conclusions on stability properties of invariant sets of dynamical systems
can be drawn without explicit knowledge of solutions and solely based on the time
derivative of a Lyapunov function along solutions. While it is frequently nontrivial to
find appropriate Lyapunov functions, in many cases it is significantly less challenging
than the derivation of an explicit solution of the dynamical system.

Although Lyapunov functions were originally proposed by Lyapunov to provide
sufficient conditions for stability properties of equilibria of differential equations,
subsequent works by Barbashin and Krasovskii [8], Malkin [52], Massera [53], and

Chetaev [19], for example, extended the results to other types of stability, applied them to more general dynamical systems, and began to address the converse question or the necessity of the various Lyapunov conditions. For an overview of contributions and the developments of Lyapunov methods we refer to the article [37].

A reference unifying many results on stability was written by Teel and Praly [69], showing that a closed set $\mathcal{A} \subset \mathbb{R}^n$ is $\mathcal{KL}$-stable with respect to two measures and with respect to the differential inclusion

$$\dot{x} \in F(x), \qquad x_0 \in \mathbb{R}^n, \tag{1.1}$$

with upper semicontinuous right-hand side $F : \mathbb{R}^n \rightrightarrows \mathbb{R}^n$, if and only if there exists a corresponding smooth Lyapunov function. In the case that $\mathcal{A}$ represents the origin and the measures are defined as the Euclidean norm, for example, then the results reduce to a characterization of asymptotic stability. Comparison functions introduced by Massera [53] and Hahn [35] (see also [36]) have become a modern tool in stability analysis and are used in $\mathcal{KL}$-stability to replace classical asymptotic stability definitions based on ε-δ formulations and convergence.

While the results in [69] focus on properties of *all solutions* of the differential inclusion and equivalent Lyapunov characterizations, results based on the *existence of at least one solution* with specific properties is analyzed through *control Lyapunov functions*. To avoid confusion, stability properties are usually called *strong* if *all solutions* are of interest and *weak* if the *existence of a solution* is of interest.

Control Lyapunov functions originate from the works of Artstein [5] and Sontag [63]. Sontag discusses the relationship between asymptotic controllability and the existence of a continuous but possibly nonsmooth positive definite function whose derivative along the solution decreases if the input of the control system is selected appropriately. Artstein expresses similar ideas for continuously differentiable functions and more restrictive dynamics.

While concepts like strong $\mathcal{KL}$-stability and asymptotic stability are equivalent to the existence of smooth Lyapunov functions the same connection cannot be established in the context of weak $\mathcal{KL}$-stability, asymptotic controllability or stabilizability and control Lyapunov functions. Famous examples include the Brockett integrator [16] and Artstein's circles [5], which show that asymptotic controllability does not necessarily imply the existence of a smooth control Lyapunov function. To overcome limitations due to assumptions on differentiability of candidate control Lyapunov functions, nonsmooth control Lyapunov functions have been introduced using the Dini derivative or proximal gradients (see [24], for example) instead of the directional derivative. Sontag and Sussman [66] showed that asymptotic controllability or stabilizability is equivalent to the existence of a continuous control Lyapunov function, and a control Lyapunov function was used by Clarke, Ledyaev, Sontag, and Subbotin [22] to derive a stabilizing controller based on this result. In [21], Clarke, Ledyaev, Rifford and Stern proved the existence of a continuous control Lyapunov function which is locally Lipschitz continuous on a domain excluding a neighborhood around the origin. Finally, weak $\mathcal{KL}$-stability of the origin of (1.1) was shown to be equivalent to the existence of a locally Lipschitz continuous control Lyapunov function by

Rifford in [59] and by Kellett and Teel in [38, 39]. In [59] it is additionally shown that there exists a semiconcave control Lyapunov function, a property which is stronger than Lipschitz continuity, while in [39] the results are not restricted to stability of the origin, but are applicable to more general invariant sets. For semiconcavity and properties of semiconcave control Lyapunov functions we refer to the references [10, 18, 20]. Here, differences between semiconcave functions and Lipschitz continuous functions will not be addressed and we focus on stability properties of the origin instead of more general sets.

While the terminology suggests that stability is a desirable phenomenon while instability is undesirable, in practical applications this is not necessarily true. For instance, in many practical applications modeled by dynamical systems or control systems, unsafe regions of the state space exist which should be avoided by the solutions of the system. Hence, it is desirable that such unsafe sets are unstable, which requires both the analysis of instability and the design of controllers that render such sets unstable.

This motivated the writing of this monograph, whose contributions are to mirror existing results on strong and weak $\mathcal{KL}$-stability to obtain corresponding instability results and to establish the diagram in Fig. 1.1.

In particular, instead of a Lyapunov function characterizing that all solutions of (1.1) converge to the origin, we give a Lyapunov-like function guaranteeing that all solutions go to infinity for $t \to \infty$. Similarly, results on the existence of one solution with certain properties analogous to control Lyapunov functions are derived. To acknowledge the contributions of Chetaev in the context of instability of differential equations and to differentiate between stability and instability properties, in this monograph Lyapunov-like functions characterizing instability properties are called Chetaev functions and control Chetaev functions, respectively. Precise definitions are provided in due course. Moreover, connections between stability properties in forward and backward time as well as invariance with respect to (positive) scaling of differential inclusions are discussed.

While results characterizing stability are essentially complete, instability and destabilization have not been studied to the same extent. We emphasize that while it may appear that such characterizations are easily obtained via time reversal, the situation is much more subtle. On the one hand, the fact that for stability the solutions usually flow towards a compact set, while for instability the solutions flow away from a compact set causes technical difficulties. On the other hand, as we will see in Sect. 4.3 and particularly in Corollary 4.15, weak (in)stability concepts are usually characterized by nonsmooth control Lyapunov or control Chetaev functions, respectively, and the corresponding generalized differential inequalities for nonsmooth functions cannot simply be reversed in time. In [26], so called anti-control Lyapunov functions, similar to control Chetaev functions discussed here, are introduced. However, necessary and sufficient conditions for the existence of anti-control Lyapunov functions are not derived. Instead, the anti-control Lyapunov functions are used to define destabilizing feedback laws based on Sontag's universal formula [64]. Additionally, in [27], control Chetaev functions have been proposed to characterize instability properties of control systems. However, the definition of these functions

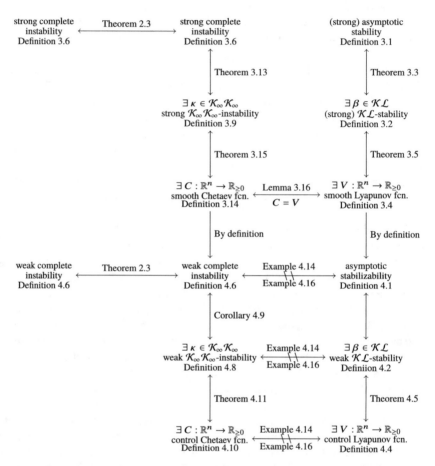

Fig. 1.1 Equivalences and differences between stability/instability properties of equilibria of differential inclusions and Lyapunov characterizations discussed in this monograph

differs from the motivation and the definition used here. Moreover, no proofs are provided in the reference [27].

Strong and weak complete instability as discussed in Chaps. 3 and 4 considers a specific type of instability which is relevant to the characterization of robust avoidance (in the strong setting) or controlled avoidance (in the weak setting), for example. In particular, strong complete instability as discussed here describes the property that all solutions—independent of the initial condition—drift away from the equilibrium. Moreover, before drifting away, solutions are only allowed to enter a certain neighborhood around the origin, whose size depends on the proximity of the initial condition to the equilibrium. Similarly, weak complete instability describes the ability to avoid a neighborhood around the origin—again independent of the initial condition—and the size of this neighborhood increases with the distance between the initial condition and the equilibrium. Other research streams discussing different concepts of instability, for example the common definition of instability as "not stable", are not discussed in this monograph.

To maintain a unified presentation, results discussed within this monograph on stability or instability are formulated with respect to the origin of a differential inclusion and are stated in terms of global results. References discussing more general cases covering forward invariant sets instead of the origin or distinguishing between local and global results are indicated within the chapters.

The results presented herein draw on a rich vein of classical results in Lyapunov theory as found in Lyapunov's original work [50], and the monographs by Hahn [35], Lakshmikantham [46], Bhatia and Szegö [9], Rouche, Habets, Laloy [60] and Filippov [30]. The survey articles [48, 49] and [37] provide an overview of several classical results, embedded in the modern control theory language. Modern monographs dedicated to Lyapunov theory include the books [7, 51, 56].

Preliminary versions of some of the results discussed in this monograph have been presented in the conference paper [11]. Similarly, first ideas combining control Lyapunov functions and control Chetaev functions to guarantee stability and avoidance properties simultaneously were presented in the conference papers [13, 14] and are outlined in Chap. 5 of this monograph. The results on combined stabilization and destabilization motivate the results on Chetaev and control Chetaev functions in the first place. Barrier functions and control barrier functions address a related problem and are briefly discussed and put in context to the material of this monograph in Sect. 4.4. However, a detailed treatment of these functions is out of the scope of this monograph and we refer to [1, 2], for example, for more information.

The monograph is structured as follows. In Chap. 2 the mathematical setting is introduced and necessary tools in terms of differential inclusions and Dini derivatives are motivated.

In Chap. 3 existing results on equivalences between asymptotic stability and the existence of smooth Lyapunov functions are reviewed. Afterwards the results are mirrored to characterize *strong complete instability* or equivalently strong $\mathcal{K}_\infty \mathcal{K}_\infty$-instability through *smooth Chetaev functions*. In particular, following the arguments in [69], equivalences between strong $\mathcal{K}_\infty \mathcal{K}_\infty$-instability of the origin of (1.1) and the existence of a smooth Chetaev functions are discussed. Moreover, it is shown that strong $\mathcal{K}\mathcal{L}$-stability in forward time is equivalent to strong $\mathcal{K}_\infty \mathcal{K}_\infty$-instability in backward time and it is proven that stability properties are invariant under a positive scaling.

In Chap. 4 the results on Lyapunov and Chetaev functions are extended to control Lyapunov and Control Chetaev functions. Here the main result shows that weak $\mathcal{K}_\infty \mathcal{K}_\infty$-instability is equivalent to the existence of a continuous control Chetaev function similar to the equivalence between weak $\mathcal{K}\mathcal{L}$-stability and the existence of a Lipschitz continuous control Lyapunov function. Additionally, by discussing appropriate examples, we prove that weak stability in forward time does not necessarily guarantee weak instability in backward time and vice versa.

In Chap. 5 ideas on merging control Lyapunov and control Chetaev functions in the controller design are discussed to guarantee combined stability and instability, i.e., avoidance, properties. In particular, two approaches are considered. The first approach combines control Lyapunov and control Chetaev functions into a single framework called complete control Lyapunov function. The second approach uses a hybrid systems framework to switch between stabilizing and destabilizing control laws designed based on control Lyapunov and control Chetaev functions and guaranteeing convergence and avoidance depending on the proximity of the state to the origin or a potential obstacle to be avoided.

Proofs of the main results are deferred to Chap. 6, in order to avoid a disruption of the flow of presentation. Auxiliary results needed for the proofs in Chap. 6 are collected in Chap. 7.

Notation

Throughout this monograph the following notation and definitions are used. For two sets $\mathcal{A}, \mathcal{B} \subset \mathbb{R}^n$, $\mathcal{A} + \mathcal{B}$ denotes the Minkowski sum, i.e., $\mathcal{A} + \mathcal{B} = \{a + b | a \in \mathcal{A}, b \in \mathcal{B}\}$. Moreover, for $r \in \mathbb{R}_{\geq 0}$ and $\mathcal{A} \subset \mathbb{R}^n$, the set $r\mathcal{A}$ is defined as $r\mathcal{A} = \{ra \in \mathbb{R}^n | a \in \mathcal{A}\}$. The closure of a set $\mathcal{A} \subset \mathbb{R}^n$ is denoted by $\overline{\mathcal{A}}$ and $\overline{\mathrm{conv}}(\mathcal{A})$ denotes the closure of its convex hull. The norm of a vector $x \in \mathbb{R}^n$ is denoted by $|x|$. For $x \in \mathbb{R}^n$ we use $B_\varepsilon(x) = \{y \in \mathbb{R}^n | |x - y| < \varepsilon\}$ to denote the open ball of radius $\varepsilon > 0$, centered around x. Note that through the earlier definition $B_\varepsilon(0)$ coincides with $\varepsilon B_1(0)$.

The stability results and definitions are based on so-called comparison functions. A continuous function $\rho : \mathbb{R}_{\geq 0} \to \mathbb{R}_{\geq 0}$ is said to be of class $\mathcal{P}$ ($\rho \in \mathcal{P}$) if $\rho(0) = 0$, and $\rho(s) > 0$ for all $s > 0$. A function $\alpha \in \mathcal{P}$ is said to be of class $\mathcal{K}$ ($\alpha \in \mathcal{K}$) if it is strictly increasing. A function $\alpha \in \mathcal{K}$ is said to be of class $\mathcal{K}_\infty$ ($\alpha \in \mathcal{K}_\infty$) if $\lim_{s \to \infty} \alpha(s) = \infty$. A continuous function $\sigma : \mathbb{R}_{\geq 0} \to \mathbb{R}_{\geq 0}$ is said to be of class $\mathcal{L}$ ($\sigma \in \mathcal{L}$), if it is strictly decreasing, and $\lim_{s \to \infty} \sigma(s) = 0$. A continuous function $\beta : \mathbb{R}_{\geq 0}^2 \to \mathbb{R}_{\geq 0}$ is said to be of class $\mathcal{KL}$ ($\beta \in \mathcal{KL}$) if $\beta(\cdot, s) \in \mathcal{K}_\infty$ for all $s \in \mathbb{R}_{\geq 0}$ and $\beta(s, \cdot) \in \mathcal{L}$ for all $s \in \mathbb{R}_{\geq 0}$.

Chapter 2
Mathematical Setting and Motivation

Abstract In this chapter we introduce the general setting in terms of differential inclusions and their solutions. Additionally, we recall stability notions for differential equations and corresponding Lyapunov-like characterizations. Since in the context of differential inclusions smooth control Lyapunov functions are not sufficient to describe weak stability properties, we use nonsmooth control Lyapunov functions in the Dini sense. Nonsmooth control Lyapunov functions of this form were introduced by Sontag in [63]. The Dini derivative as an extension of the directional derivative is discussed in the last part of this chapter.

Keywords Lyapunov methods · Differential inclusions · Stability of nonlinear systems · Instability of nonlinear systems · Stabilization/destabilization of nonlinear systems · Stabilizability and destabilizability

2.1 Differential Inclusions

We consider dynamical systems described through differential inclusions

$$\dot{x} \in F(x), \qquad x_0 \in \mathbb{R}^n, \tag{2.1}$$

for a set-valued map $F : \mathbb{R}^n \rightrightarrows \mathbb{R}^n$, and an initial value $x(0) = x_0 \in \mathbb{R}^n$. We are interested in stability properties of the origin and assume that without loss of generality $0 \in F(0)$ holds. While the origin can be replaced by more general sets, in the results presented in the following we restrict our attention to the equilibrium $x^e = 0$ for simplicity. To guarantee existence of solutions of (2.1) we will make use of the following assumption on F throughout the monograph.

Assumption 2.1 (Basic conditions) The set-valued map $F : \mathbb{R}^n \rightrightarrows \mathbb{R}^n$ with $0 \in F(0)$ has nonempty, compact, and convex values on $\mathbb{R}^n$, and it is upper semicontinuous.

© The Author(s), under exclusive license to Springer Nature Switzerland AG 2021
P. Braun et al., *(In-)Stability of Differential Inclusions*,
SpringerBriefs in Mathematics,
https://doi.org/10.1007/978-3-030-76317-6_2

Here the compact set-valued map $F : \mathbb{R}^n \rightrightarrows \mathbb{R}^n$ is upper semicontinuous if for each $x \in \mathbb{R}^n$ and for all $\varepsilon > 0$ there exists a $\delta > 0$ such that for all $\xi \in B_\delta(x)$ we have $F(\xi) \subset F(x) + B_\varepsilon(0)$.

For some results Assumption 2.1 is not sufficient and additionally Lipschitz continuity of F is necessary. We recall that F is Lipschitz continuous on $\mathbb{R}^n \backslash \{0\}$ if there exists a constant $L > 0$ and a neighborhood $O \subset \mathbb{R}^n$ of $x \in \mathbb{R}^n \backslash \{0\}$ such that

$$F(x_1) \subset F(x_2) + B_{L|x_1 - x_2|}(0),$$

for all $x_1, x_2 \in O$, [6, Def. 1.4.5].

Assumption 2.2 (Lipschitz continuity) The set-valued map $F : \mathbb{R}^n \rightrightarrows \mathbb{R}^n$ with $0 \in F(0)$ is locally Lipschitz continuous on $\mathbb{R}^n \backslash \{0\}$.

Assumption 2.1 ensures that absolutely continuous solutions $\phi(\cdot; x_0) : [0, T) \to \mathbb{R}^n$, $(T \in \mathbb{R}_{>0} \cup \{\infty\})$ satisfying the differential inclusion (2.1) for almost all $t \in [0, T)$ exist for any initial value $x_0 \in \mathbb{R}^n$. In particular, these solutions are differentiable almost everywhere. The set of all such solutions $\phi(\cdot; x_0)$ with $\phi(0; x_0) = x_0$ is denoted by $S(x_0)$.

Solutions $\phi(\cdot; x_0)$ are finite on a maximal time interval. To simplify the exposition in the following, we define solutions $\phi(\cdot; x_0) : \mathbb{R} \to \mathbb{R}^n \cup \{\pm\infty\}^n$ as extended real valued functions. In this case $\phi(\cdot; x_0)$ is defined for all $t \in \mathbb{R}$ even in the case of finite escape time. Additionally, for the ith component of the n-vector solution $\phi(\cdot; x_0)$ we will use the following convention:

- If $\phi_i(T; x_0) = \pm\infty$ for $T > 0$ and $i \in \{1, \ldots, n\}$, then $\phi_i(t; x_0) = \pm\infty$ for all $t \geq T$.
- If $\phi_i(T; x_0) = \pm\infty$ for $T < 0$ and $i \in \{1, \ldots, n\}$, then $\phi_i(t; x_0) = \pm\infty$ for all $t \leq T$.

Solutions which satisfy $|\phi(t; x_0)| < \infty$ for all $t \in \mathbb{R}_{\geq 0}$ are called forward complete.

In the following chapters we discuss stability and instability results for differential inclusions (2.1) in forward time, $t \to \infty$. However, based on these results, we additionally discuss implications for time reversal solutions, $t \to -\infty$. An extended real valued function $\psi(\cdot; x_0) : \mathbb{R} \to \mathbb{R}^n \cup \{\pm\infty\}^n$ is called a time reversal solution of the differential inclusion (2.1) if

$$\psi(t; x_0) = \phi(-t; x_0)$$

for a $\phi(\cdot; x_0) \in S(x_0)$ for all $t \in \mathbb{R}$. A time reversal solution satisfies the differential inclusion

$$\dot{x} \in -F(x), \qquad x_0 \in \mathbb{R}^n, \tag{2.2}$$

for almost all $t \in \mathbb{R}$.

The differential inclusion (2.2) represents a particular scaling of the original differential inclusion (2.1). To avoid finite escape time we make use of another common scaling of (2.1) which ensures that all solutions of the scaled differential inclusion are forward and backward complete while preserving the stability properties of the original differential inclusion. This concept is used, for example, in the publications [17, 23, 47, 59, 69] and [30, Chap. 2, Sect. 9] with similar motivations. More precisely, for a positive continuous function $\eta : \mathbb{R}_{\geq 0} \to \mathbb{R}_{>0}$, we consider the scaled differential inclusion

$$\dot{x} \in F_\eta(x) = \eta(|x|) F(x), \qquad x_0 \in \mathbb{R}^n. \tag{2.3}$$

Note that in contrast to the definition of positive definite functions, for positive functions we demand that $\eta(0) > 0$ instead of $\eta(0) = 0$. A negative scaling is achieved by scaling the differential inclusion (2.2). Additionally observe that for continuous positive functions η, the differential inclusion (2.3) satisfies the basic conditions (Assumption 2.1) if F satisfies Assumption 2.1. This fact follows immediately from the properties of η and the properties of F. The set of solutions of (2.3) is denoted by $\mathcal{S}_\eta(\cdot)$.

Theorem 2.3 (Positive scaling of differential inclusions) *Consider the differential inclusion (2.1) satisfying Assumption 2.1. Let $\eta : \mathbb{R}_{\geq 0} \to \mathbb{R}_{>0}$ be a continuous positive function corresponding to the scaled differential inclusion (2.3) and denote the solution set of (2.3) from $x_0 \in \mathbb{R}^n$ by $\mathcal{S}_\eta(x_0)$.*
For all $x_0 \in \mathbb{R}^n$ and for all $\phi(\cdot; x_0) \in \mathcal{S}(x_0)$ with

$$|\phi(t; x_0)| < \infty, \quad \forall\, t < T \qquad and \qquad |\phi(t; x_0)| = \infty \;\; \forall\, t \geq T,$$

$T \in \mathbb{R}_{>0} \cup \{\infty\}$, there exist a continuous strictly increasing function $\alpha : [0, T) \to [0, M)$ and $M \in \mathbb{R}_{>0} \cup \{\infty\}$ with $\alpha(0) = 0$ such that $\phi_\eta(\cdot; x_0) = \phi(\alpha(\cdot); x_0) \in \mathcal{S}_\eta(x_0)$. Conversely, if $\phi_\eta(\cdot; x_0) \in \mathcal{S}_\eta(x_0)$ then $\phi_\eta(\alpha^{-1}(\cdot); x_0) \in \mathcal{S}(x_0)$ is satisfied. Moreover, in the limit, the solutions satisfy

$$\lim_{t \to T} |\phi(t; x_0)| = \lim_{t \to M} |\phi_\eta(t; x_0)|. \tag{2.4}$$

Before we prove this result at the end of this section, we discuss its implications. Theorem 2.3 essentially states that solutions of (2.1) and (2.3) have the same "shape" and only differ in their "velocity" captured through the time argument t and $\alpha(t)$, respectively.

If $T = M = \infty$ (i.e., $\alpha \in \mathcal{K}_\infty$), then a forward complete solution $\phi(\cdot; x_0) : \mathbb{R}_{\geq 0} \to \mathbb{R}^n$ is mapped to a forward complete solution $\phi(\alpha(\cdot); x_0) : \mathbb{R}_{\geq 0} \to \mathbb{R}^n$ while preserving the asymptotic behavior. However, the result is more general and also allows that T and/or M are finite while solutions preserve their properties for $t \to T$ and $t \to M$, respectively.

Here, we consider a particular scaling η which ensures that all solutions of (2.3) are forward complete independent of the properties of the solutions of (2.1). In particular we define

$$\tilde{v}(r) = \max_{y \in F(x), |x|=r} |y| \tag{2.5}$$

and $v : \mathbb{R}_{\geq 0} \rightarrow \mathbb{R}_{\geq 0}$ continuous such that $v(r) \geq \tilde{v}(r)$ for all $r \geq 0$. Then η defined as

$$\eta(r) = \frac{1}{v(r) + 1}. \tag{2.6}$$

satisfies $\eta(r) \subset B_1(0)$ for all $r \geq 0$.

Corollary 2.4 *Consider the differential inclusion* (2.1) *satisfying Assumption 2.1. Then there exists a continuous positive function* $\eta : \mathbb{R}_{\geq 0} \rightarrow \mathbb{R}_{>0}$ *such that*

$$\eta(|x|)F(x) \subset \overline{B}_1(0)$$

for all $x \in \mathbb{R}^n$. *Moreover the set-valued map* $\eta(|\cdot|)F(\cdot) : \mathbb{R}^n \rightrightarrows \mathbb{R}^n$ *satisfies Assumption 2.1 and all solutions of* (2.3) *are forward complete.*

Proof Since $F(x)$ and $\overline{B}_{|x|}(0)$ are compact sets for all $x \in \mathbb{R}^n$, the function $\tilde{v} : \mathbb{R}_{\geq 0} \rightarrow \mathbb{R}_{\geq 0}$ defined in (2.5) is well defined. Then, using the function η in (2.6), for all $x \in \mathbb{R}^n$ and for all $y \in F(x)$, it holds that

$$\frac{1}{v(|x|) + 1} y \leq \frac{1}{\tilde{v}(|x|) + 1} y \in B_1(0),$$

or equivalently for all $x \in \mathbb{R}^n$ the inclusions

$$\frac{1}{v(|x|) + 1} F(x) \subset \frac{1}{\tilde{v}(|x|) + 1} F(x) \subset B_1(0)$$

are satisfied. Since η is continuous by construction, $\eta(\cdot)F(\cdot)$ is upper semicontinuous. Forward completeness of solutions of 2.3 follows from [68, Corollary 1] or [61, Theorem 3.3], for example. □

In [47], the scaling is discussed in Lemmas 7.1 and 7.2, where Lemma 7.1 guarantees the existence of an appropriate scaling while Lemma 7.2 shows that stability properties under scaling are preserved. In [23], the scaled system (2.3) is covered in Lemma 2.4. In [17], the scaling is applied to control systems to ensure forward completeness of solutions while preserving properties of the original control system. In [59], Propositions 1 and 2 are used to ensure that the set-valued map $\eta(|x|)F(x)$ is uniformly bounded. Combining this results with [59, Assumption 3.2] guarantees that solutions of (2.3) are forward complete in the same way as argued here. In [69], scaling is used to establish and characterize backward completability with respect to a measure and subsequently to establish robust stability properties of a differential inclusion. In [30, Chap. 2, Sect. 9] the scaling is discussed in the general context of change of variables. We refer to [69, Definition 9], [69, Proposition 2], [69, Theorem 3] and [30, Chap. 2, Sect. 9, Theorem 3] for details.

Example 2.5 Consider the differential equation

$$\dot{x} = x^3$$

as a special case of the differential inclusion (2.1) satisfying Assumption 2.1. The solution of the differential equation is given by

$$\phi(t; x_0) = \frac{x_0}{\sqrt{1 - 2t x_0^2}}$$

and we can observe finite escape time. In particular for $x_0 = 1$ it holds that $\phi(\cdot, 1) :$ $[0, \frac{1}{2}) \to \mathbb{R}$ and $\lim_{t \to \frac{1}{2}} \phi(t, 1) = \infty$.

Consider the continuous positive function

$$\eta(r) = \begin{cases} 1 & \text{for } r \le 1 \\ \frac{1}{r^3} & \text{for } r \ge 1 \end{cases}$$

leading to the scaled differential inclusion $\dot{y} = \eta(|y|) y^3$ defined as

$$\dot{y} = \begin{cases} -1, & \text{for } y \in (-\infty, -1], \\ y^3, & \text{for } y \in [-1, 1], \\ 1, & \text{for } y \in [1, \infty). \end{cases}$$

For $y_0 = 1$, it holds that

$$\phi_\eta(t; 1) = 1 + t$$

and thus $\phi_\eta(\cdot; 1)$ is finite for all $t \in \mathbb{R}_{\ge 0}$ with limit $\lim_{t \to \infty} \phi_\eta(t, 1) = \infty$.

According to Theorem 2.3 there exists a continuous strictly increasing function $\alpha : \mathbb{R}_{\ge 0} \to [0, \frac{1}{2}]$ such that $\phi(\alpha(t), 1) = \phi_\eta(t, 1)$ for all $t \in \mathbb{R}_{\ge 0}$. Indeed, using this condition it holds that

$$\phi(\alpha(t), 1) = \frac{1}{\sqrt{1 - 2\alpha(t)}} = t + 1 = \phi_\eta(t, 1)$$

which implies that

$$\alpha(t) = \frac{1}{2} - \frac{1}{2} \left(\frac{1}{1+t} \right)^2$$

and α satisfies the conditions of Theorem 2.3.

Before starting with the proof of Theorem 2.3, we give the following lemma as an auxiliary result.

Lemma 2.6 *Let $\psi : [0, M) \to \mathbb{R}^n$, $M \in \mathbb{R}_{>0} \cup \{\infty\}$, be an absolutely continuous function. Additionally, let $\eta : \mathbb{R}_{\ge 0} \to \mathbb{R}_{>0}$ be a continuous positive function. Con-*

sider the one dimensional initial value problem

$$\dot\alpha(t) = \eta(|\psi(\alpha(t))|), \qquad \alpha(0) = 0.$$

Then there exists $T \in \mathbb{R}_{>0} \cup \{\infty\}$ such that $\alpha : [0, T) \to [0, \infty)$ is a unique continuously differentiable solution of the initial value problem, which is strictly monotonically increasing, $\alpha(0) = 0$ and $\lim_{t \to T} \alpha(t) = \infty$ if $T < \infty$.

Proof Since $\eta(|\psi(\cdot)|)$ is continuous, Peano's existence theorem guarantees a local solution of the initial value problem on $[0, T)$, $T > 0$, satisfying $\alpha(0) = 0$. The fact that $\eta(|\psi(s)|) > 0$ for all $s \in \mathbb{R}_{\geq 0}$ implies that $\alpha(\cdot)$ is strictly monotonically increasing. Let $[0, T) \subset \mathbb{R}_{\geq 0} \cup \{\infty\}$ be the maximal interval of existence of the solution. If $T < \infty$, then $\lim_{t \to T} \alpha(t) = \infty$, since otherwise the interval of existence can be extended.   □

Proof of Theorem 2.3 Let $\phi \in S(x_0)$, $x_0 \in \mathbb{R}^n$ be an arbitrary solution of (2.1). Let $\alpha(t)$ be the solution of the initial value problem

$$\dot\alpha(t) = \eta(|\phi(\alpha(t), x_0)|), \qquad \alpha(0) = 0. \tag{2.7}$$

We denote the domain of α by $[0, T)$, $T \in \mathbb{R}_{>0} \cup \{\infty\}$ and the image is defined as $[0, M)$, where $M = \lim_{t \to T} \alpha(t)$.

We define the coordinate transformation $\tilde{t} = \alpha(t)$. Then for almost all $\tilde{t} \in [0, M)$ it holds that

$$\frac{d}{dt}\phi(\alpha(t); x_0) = \frac{d}{dt}\alpha(t)\frac{d}{d\tilde{t}}\phi(\tilde{t}; x_0) = \eta(|\phi(\alpha(t); x_0)|)\frac{d}{d\tilde{t}}\phi(\tilde{t}; x_0)$$
$$= \eta(|\phi(\tilde{t}; x_0)|)\frac{d}{d\tilde{t}}\phi(\tilde{t}; x_0).$$

Thus $\phi(\tilde{t}, x_0) = \phi(\alpha(t); x_0)$ satisfies the scaled differential inclusion (2.3) for almost all $\tilde{t} \in [0, M)$ and thus $\phi(\alpha(\cdot); x_0) \in S_\eta(x_0)$.

Vice versa, if we assume that $\phi(\cdot; x_0) \in S_\eta(x_0)$ we need to show that $\phi(\alpha^{-1}(\cdot); x_0) \in S(x_0)$. It holds that

$$\frac{d}{d\tilde{t}}\phi(\alpha^{-1}(\tilde{t}); x_0) = \frac{d}{dt}\left(\alpha^{-1}(\tilde{t})\right)\frac{d}{dt}\phi(t; x_0) = \frac{1}{\dot\alpha(\alpha^{-1}(\tilde{t}))}\eta(|\phi(t; x_0)|)F(\phi(t; x_0))$$

$$= \frac{1}{\eta(|\phi(t; x_0)|)}\eta(|\phi(t; x_0)|)\frac{d}{dt}\phi(t; x_0) = F(\phi(t; x_0))$$

which shows the other direction.

Since the solutions coincide up to the time scaling, the property (2.4) follows. □

In the context of ordinary differential equations a similar argument with respect to scaling is made in [9, Theorem 3.1.67], for example. Similar to the use here, in [9] it is argued that forward completeness of solutions can be assumed without loss of generality when it comes to qualitative properties such as stability and instability of equilibria of differential equations.

As an example of the differential inclusion (2.1) and as an important motivation for the stability notions discussed in this monograph, we consider control systems

$$\dot{x} = f(x, u), \qquad x_0 \in \mathbb{R}^n, \qquad u \in \mathcal{U}(x) \subset \mathbb{R}^m \tag{2.8}$$

where x denotes the state and u either represents an input or a disturbance.

In this case, the set-valued map F in (2.1) can be defined as

$$F(x) = \overline{\text{conv}}\{f(x, u) \in \mathbb{R}^n | u \in \mathcal{U}(x)\} \tag{2.9}$$

and $\overline{\text{conv}}(\cdot)$ denotes the closure of the convex hull. Sufficient conditions for F in (2.9) to satisfy Assumptions 2.1 and 2.2 are that $f : \mathbb{R}^n \times \mathbb{R}^m \to \mathbb{R}^n$ is locally Lipschitz continuous in $x \in \mathbb{R}^n$ and continuous in $u \in \mathbb{R}^m$ and that $\mathcal{U} = \mathcal{U}(x)$ for all $x \in \mathbb{R}^n$ is compact or that $\mathcal{U}(x) = B_{c|x|}(0)$ for $c > 0$, [39, Remark 4].

In the case that u represents a disturbance, Chap. 3 is dedicated to the question of the qualitative behavior of all solutions of the system (2.8). In particular, we will investigate under which conditions all solutions—i.e., independent of the selection of the disturbance $u : \mathbb{R}_{\geq 0} \to \mathcal{U}(x)$ and the initial state $x_0 \in \mathbb{R}^n$—converge to the origin. In this case, the convergence behavior is robust with respect to u and x_0. This setting will be discussed in Sect. 3.1. Similarly, we will study necessary and sufficient conditions to guarantee that all solutions starting arbitrarily close to the origin eventually leave every bounded set around the origin. This is discussed in Sect. 3.2.

If u represents an input, i.e., u can be actively selected, stabilizability and desta-bilizability will be discussed in Sects. 4.1 and 4.2, respectively. Here, we will discuss characterizations which guarantee that every initial condition can be asymptotically driven to the origin or that every domain around the origin can be left by at least one solution, i.e., if u is selected appropriately.

However, before we turn to the analysis of the differential inclusion (2.1) we discuss the special case of differential equations, where, due to uniqueness, the cases "for all solutions" and "there exists a solution" coincide.

2.2 (In)stability Characterizations for Ordinary Differential Equations

Before we discuss stability and instability of the origin of differential inclusions we recall stability notions and corresponding Lyapunov characterizations for differential equations. We consider ordinary differential equations

$$\dot{x} = f(x), \qquad x_0 \in \mathbb{R}^n, \tag{2.10}$$

as a special case of the differential inclusion (2.1), with a Lipschitz continuous right-hand side $f : \mathbb{R}^n \to \mathbb{R}^n$. Again, we assume that $f(0) = 0$ is satisfied. In contrast to the generalized definition (2.1), solutions of (2.10) are unique and $\mathcal{S}(x_0)$ contains only a single element for each $x_0 \in \mathbb{R}^n$.

Definition 2.7 ((*Global*) *stability*) The origin is (Lyapunov) stable for (2.10) if, there exists $\delta \in \mathcal{K}_\infty$ such that for all $\varepsilon \geq 0$,

$$|\phi(t; x_0)| \leq \varepsilon \qquad \text{whenever } |x_0| \leq \delta(\varepsilon) \text{ and } t \geq 0. \tag{2.11}$$

Since we require that (2.11) hold for all $\varepsilon \geq 0$, the property defined in Definition 2.7 is global. If instead we demand the existence of $\bar{\varepsilon} > 0$ such that (2.11) holds for all $\varepsilon \in [0, \bar{\varepsilon}]$, then local stability of the origin is obtained. As highlighted in the introduction, to simplify and to unify the presentation, if not stated explicitly, only global results are discussed and presented in this monograph, despite the existence of equivalent local results. With this in mind, the term global is usually dropped in the following.

A sufficient condition for the stability of equilibria is the existence of a Lyapunov function.

Theorem 2.8 (Lyapunov stability theorem) *Given (2.10) with $f(0) = 0$, suppose there exist a smooth function $V : \mathbb{R}^n \to \mathbb{R}_{\geq 0}$ and functions $\alpha_1, \alpha_2 \in \mathcal{K}_\infty$ such that, for all $x \in \mathbb{R}^n$,*

$$\alpha_1(|x|) \leq V(x) \leq \alpha_2(|x|),$$
$$\langle \nabla V(x), f(x) \rangle \leq 0.$$

Then the origin is (globally) stable.

The function V in Theorem 2.8 is called Lyapunov function. A stronger property of the equilibrium defined next, can similarly be concluded from appropriate Lyapunov functions.

Definition 2.9 ((*Global*) *asymptotic stability*) The origin is asymptotically stable for (2.10) if it is stable (as per Definition 2.7) and if for all $x_0 \in \mathbb{R}^n$,

$$|\phi(t; x_0)| \to 0 \qquad \text{for } t \to \infty.$$

Theorem 2.10 (Lyapunov asymptotic stability theorem) *Given (2.10) suppose there exist a smooth function $V : \mathbb{R}^n \to \mathbb{R}_{\geq 0}$, functions $\alpha_1, \alpha_2 \in \mathcal{K}_\infty$, and a positive definite function $\rho \in \mathcal{P}$ such that, for all $x \in \mathbb{R}^n$*

$$\alpha_1(|x|) \leq V(x) \leq \alpha_2(|x|), \tag{2.12}$$
$$\langle \nabla V(x), f(x) \rangle \leq -\rho(|x|). \tag{2.13}$$

Then the origin is (globally) asymptotically stable.

An equilibrium which is not stable is called unstable.

Definition 2.11 (*Instability*) The origin is unstable for system (2.10) if it is not stable.

Since instability is simply defined as not stable, there are many definitions distinguishing between different instability properties. For example for differential equations where the role of the origin is replaced by a more general set, Bhatia and Szegö discuss and distinguish between *instability*, *weak instability* (see [9, Definition 1.6.34]), *ultimate instability*, *ultimate weak instability* (see [9, Definition 1.6.37]), *weak complete instability*, *ultimate weak complete instability*, *complete instability* and *ultimate complete instability* (see [9, Definition 1.6.39]).

In this monograph we focus on *complete instability* as defined next, mirroring Definition 2.9 in terms of the asymptotic behavior for $t \to \infty$.

Definition 2.12 (*(Global) complete instability*) The origin is completely unstable for (2.10) if there exists $\alpha \in \mathcal{K}_\infty$ such that for all $\delta > 0$ the condition $x_0 \in \mathbb{R}^n \setminus B_{\alpha(\delta)}(0)$ implies

$$|\phi(t; x_0)| \geq \delta \qquad\qquad \forall\, t \in \mathbb{R}_{\geq 0}, \qquad (2.14)$$
$$|\phi(t; x_0)| \to \infty \qquad\qquad \text{for } t \to \infty. \qquad (2.15)$$

Note that the definition of complete instability used here does not coincide with [9, Definition 1.6.39]. In [9, Definition 1.6.39], complete instability is defined through the condition that solutions starting arbitrarily close to the origin (or a more general forward invariant set) need to leave an ε-neighborhood around the origin at some time $t \in \mathbb{R}_{\geq 0}$ but may reenter the ε-neighborhood for $t \to \infty$. A similar local definition of complete instability is given in [34, Definition 2.6]. A modern reference covering complete instability is [54], (see in particular [54, Definition 3.1.20]).

Complete instability can be similarly concluded using Lyapunov-like arguments, where the decrease condition (2.13) is replaced by an increase condition.

Theorem 2.13 (Lyapunov complete instability theorem) *Consider* (2.10) *and suppose there exist a smooth function* $C : \mathbb{R}^n \to \mathbb{R}_{\geq 0}$, *functions* $\alpha_1, \alpha_2 \in \mathcal{K}_\infty$, *and a positive definite function* $\rho \in \mathcal{P}$ *such that, for all* $x \in \mathbb{R}^n$,

$$\alpha_1(|x|) \leq C(x) \leq \alpha_2(|x|),$$
$$\langle \nabla C(x), f(x) \rangle \geq \rho(|x|).$$

Then the origin is (globally) completely unstable.

For proofs of Theorems 2.8, 2.10, and 2.13 we refer to the monographs [35, 40, 54]. However, these theorems also follow as special cases of the results discussed in Chap. 3.

For instability (in contrast to complete instability) the conditions on the functions C in Theorem 2.13 can be relaxed.

Theorem 2.14 (Chetaev's theorem, [40, Theorem 4.3]) *Consider* (2.10) *and assume there exists a smooth function* $C : \mathbb{R}^n \to \mathbb{R}$ *with* $C(0) = 0$ *and* $O_r = \{x \in B_r(0) : V(x) > 0\} \neq \emptyset$ *for all* $r > 0$. *If for certain* $r > 0$,

$$\langle \nabla C(x), f(x) \rangle > 0 \quad \forall x \in O_r$$

then the origin is unstable.

Theorem 2.14 dates back to the work by Chetaev, [19]. To distinguish between stability and instability results, we call functions used with stability properties, as in Theorems 2.8 and 2.10 *Lyapunov functions* while we call functions used with instability properties, as in Theorems 2.13 and 2.14, *Chetaev functions*.

The conditions in Theorems 2.10, and 2.13 are not only sufficient but also necessary. In particular, if the origin of (2.10) is asymptotically stable or completely unstable, then Lyapunov or Chetaev functions exist satisfying the properties in Theorems 2.10 and 2.13, respectively. Theorem 2.8 is only sufficient. To obtain a converse result, more general definitions relying, for instance, on discontinuous Lyapunov functions are necessary (see [7, Theorem 2.5], [9, Theorem 1.71] for a necessary and sufficient result and [44], [7, Example 2.2] for a counterexample). In the case of Theorem 2.14, under an additional assumption (see Property A in [43, Definition 4.1]), not only does a Chetaev function as in the theorem exist when the origin is unstable, but a Chetaev function exists where the set O_r coincides with the so-called region of instability (see [37, Theorem 8.1] or [43, Theorem 7.1]). With these equivalences, we frequently refer to Lyapunov or Chetaev functions as *characterizations* of the respective stability or instability properties.

Remark 2.15 Note that, as stated, the definition and characterizations are essentially global as they are stated for all all $x \in \mathbb{R}^n$ and for all $\varepsilon > 0$. Local versions are easily obtained by restricting ε and by restricting the attention to a domain around the origin.

To illustrate complete instability, instability, and asymptotic stability and their Lyapunov characterizations we discuss simple examples using linear differential equations.

Example 2.16 Consider the three linear differential equations and their solutions

$$f_1(x) = \begin{bmatrix} x_1 \\ x_2 \end{bmatrix}, \qquad \phi_1(t; x_0) = \begin{bmatrix} x_{1,0}e^t \\ x_{2,0}e^t \end{bmatrix}, \qquad (2.16a)$$

$$f_2(x) = \begin{bmatrix} -x_1 \\ x_2 \end{bmatrix}, \qquad \phi_2(t; x_0) = \begin{bmatrix} x_{1,0}e^{-t} \\ x_{2,0}e^t \end{bmatrix}, \qquad (2.16b)$$

$$f_3(x) = \begin{bmatrix} -x_1 \\ -x_2 \end{bmatrix}, \qquad \phi_3(t; x_0) = \begin{bmatrix} x_{1,0}e^{-t} \\ x_{2,0}e^{-t} \end{bmatrix}. \qquad (2.16c)$$

For the ordinary differential equation (2.16a) all solutions $\phi_1(\cdot; x_0)$, $x_0 \neq 0$ satisfy $|\phi_1(t; x_0)| \to \infty$ for $t \to \infty$, which indicates that the origin is completely unstable.

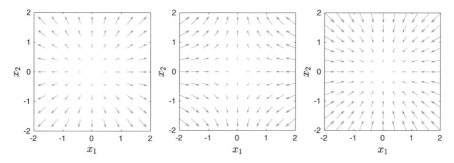

Fig. 2.1 Phase portraits of the differential equations (2.16) from the left to right. The stability properties of the origin can be intuitively understood from the arrows pointing towards or away from the origin

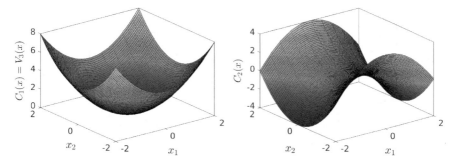

Fig. 2.2 Visualization of possible Lyapunov and Chetaev functions for the linear differential equations (2.16)

Similarly, for (2.16c) all solutions satisfy $|\phi_3(t; x_0)| \to 0$ for $t \to \infty$ for all $x_0 \in \mathbb{R}^n$, showing asymptotic stability of the equilibrium.

For (2.16b), using the initial values $x_0 = [0, c]^T$, $c \neq 0$, one obtains $|\phi_2(t; x_0)| \to \infty$ for $t \to \infty$, while for the initial values $x_0 = [c, 0]^T$, $c \neq 0$, it follows that $|\phi_2(t; x_0)| \to 0$ for $t \to \infty$. This means that the origin $x = 0$ is unstable but not completely unstable. These properties can also be rigorously verified using the Chetaev function $C_1(x) = x^T x$ and Theorem 2.13, the Chetaev function $C_2(x) = -x_1^2 + x_2^2$ and Theorem 2.14, and the Lyapunov function $V_3(x) = x^T x$ and Theorem 2.10, respectively. Fig. 2.1 shows the phase portrait of the ordinary differential equations and the Lyapunov and Chetaev functions are illustrated in Fig. 2.2.

Note that Lyapunov himself studied instability in his seminal contribution [50]. In particular, Lyapunov's first theorem on instability [50, Sect. 16, Theorem II] covers instability of the origin of (2.16a) and the function $C_1(x) = x^T x$ can be understood as a Lyapunov function for instability. However, as motivated in the introduction, we deviate from this terminology in this monograph and instead use *Lyapunov* in the context of stability and *Chetaev* in the context of instability.

In Example 2.16, the ordinary differential equation (2.16a) is the time reversal differential equation of (2.16c) and vice versa. Recall the definition of the time reversal differential inclusion (2.2) which also holds in the special case of ordinary differential equations. While the origin of (2.16a) is completely unstable the origin of the time reversal system is asymptotically stable. This is not a coincidence for ordinary differential equations where a solution is uniquely defined through the initial condition x_0. As for example Bhatia and Szegö point out in [9, Remark 1.5.18], "[b]y reversing the direction of motion along the trajectories, sets which are completely unstable will become asymptotically stable and vice versa" and thus the origin of $\dot{x} = f(x)$ is completely unstable if and only if the origin of $\dot{x} = -f(x)$ is asymptotically stable.

A similar connection can be derived in terms of smooth Lyapunov functions and smooth Chetaev functions. From

$$\langle \nabla V(x), f(x) \rangle \leq -\rho(|x|) \qquad \Longleftrightarrow \qquad \langle \nabla V(x), -f(x) \rangle \geq \rho(|x|)$$

it follows that V is a Lyapunov function with respect to $\dot{x} = f(x)$ according to Theorem 2.10 if and only if $C = V$ is a Chetaev function with respect to $\dot{x} = -f(x)$ according to Theorem 2.13. Note that we have already observed this property in Example 2.16.

As a last point in this section we discuss condition (2.14) in Definition 2.12.

Example 2.17 Consider the two dimensional dynamics

$$\begin{aligned} \dot{x}_1 &= (c^2 - x_2^2)x_1 + x_2 \\ \dot{x}_2 &= (c^2 - x_2^2)x_2 \end{aligned} \tag{2.17}$$

with parameter $c \in \mathbb{R}_{>0}$, whose time reversal counterpart is discussed in [3, Example 2] in the context of asymptotic stability. For $x_2^2 = c^2$ the dynamics reduce to $\dot{x}_1 = x_2$ and $\dot{x}_2 = 0$. The phase portrait corresponding to the dynamics (2.17) with parameter $c = 1$ shown in Fig. 2.3 indicates that condition (2.15) is satisfied for all solutions $\phi(\cdot; x_0)$, $x_0 \neq 0$, while condition (2.14) is not satisfied.

Indeed, for all initial conditions x_0 with $x_{0,1} \in \mathbb{R}_{\geq 0}$ and $x_{0,2} = c$ there exists a time $T \in \mathbb{R}_{\geq 0}$ such that $\phi(T; x_0) = [0, c]^T$. Thus, $\alpha \in \mathcal{K}_\infty$ in Definition 2.12 cannot exist and consequently the origin is not globally completely unstable according to Definition 2.12.

The last example shows that the function $\alpha \in \mathcal{K}_\infty$ is necessary to ensure that solutions starting arbitrarily far away from the origin stay arbitrarily far away from the origin for all times $t \in \mathbb{R}_{\geq 0}$ for global complete instability. In particular, for the dynamics (2.17) with $c > 0$ small, there exist initial conditions with $|x_0|$ arbitrarily large, which pass through the neighborhood $B_{2c}(0)$ before $|\phi(t; x_0)| \to \infty$ for $t \to \infty$. Nevertheless, if we restrict our analysis of complete instability of the origin to the neighborhood $B_{\frac{1}{2}c}(0)$, i.e., instead of (2.15) we require that solutions starting in $x_0 \in B_{\frac{1}{2}c}(0)\backslash\{0\}$ leave the neighborhood $B_{\frac{1}{2}c}(0)\backslash\{0\}$ and satisfy (2.14), then the origin is locally completely unstable in Example 2.17.

Fig. 2.3 Phase portrait and several solutions of the dynamical system (2.17) with $c = 1$

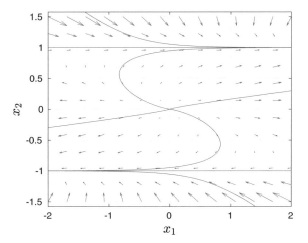

It is thus not clear to us if (2.14) is necessary for local versions of complete instability or if (2.15) implies (2.14) in this case. In the context of asymptotic stability (Definition 2.9), stability needs to be stated explicitly to ensure solutions starting in a δ-neighborhood around the origin stay in an ε-neighborhood before convergence to the origin can be observed. However, the standard example showing that attractivity is not enough to conclude asymptotic stability cannot be adapted to show the need for (2.14) in a neighborhood around the origin.

Example 2.18 The dynamical system

$$\dot{x} = f(x) = \frac{1}{|x|_2^2(1 + |x|_2^4)} \begin{bmatrix} x_1^2(x_2 - x_1) + x_2^5 \\ x_2^2(x_2 - 2x_1) \end{bmatrix} \tag{2.18}$$

introduced in [70] and further discussed in [35, Sect. 40], is known to be a system with globally attractive origin, which is not asymptotically stable.

Solutions of the differential equation (2.18) are visualized in blue together with the vector field in red. While $\phi(t; x) \to 0$ for $t \to \infty$ for all $x_0 \in \mathbb{R}^2$, the origin is not stable. In particular, there exists $\varepsilon > 0$ and sequences $(x_i)_{i \in \mathbb{N}} \subset \mathbb{R}^2$, $\lim_{i \to \infty} x_i = 0$, $(t_i)_{i \in \mathbb{N}} \subset \mathbb{R}_{\geq 0}$ such that $|\phi(t_i; x_i)| \geq \varepsilon$ for all $i \in \mathbb{N}$.

While Example 2.18 is a standard example to show that attractivity does not imply asymptotic stability, the time reversal system cannot be used to show if condition (2.14) in Definition 2.12 is necessary.

As shown in [35, Sect. 40], solutions of the dynamical systems $\dot{x} = f(x)$ and $\dot{x} = -f(x)$ starting inside the black curve visualized in Fig. 2.4 do not leave the domain enclosed by the black line. Thus, (2.14) and (2.15) are violated for the dynamics $\dot{x} = -f(x)$.

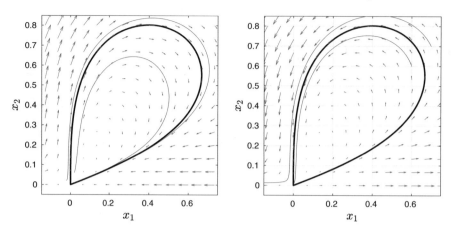

Fig. 2.4 Solutions of the dynamical systems $\dot{x} = f(x)$ (left) and $\dot{x} = -f(x)$ (right) where $f(x)$ is defined in Eq. (2.18). Solutions starting inside the black curve are trapped inside

2.3   The Dini Derivative

In contrast to differential equations, where we can rely on smooth Lyapunov and Chetaev functions, in the context of differential inclusions nonsmooth functions need to be taken into account. In this monograph we use the Dini derivative as a generalized gradient to extend results relying on smooth functions to continuous functions. Alternative extensions based on proximal subgradients or on viscosity solutions are not discussed here. For Lyapunov characterizations using proximal subgradients we refer to [20] and references therein and for viscosity solutions to [17] and references therein. The survey paper [58] reviews nonsmooth Lyapunov functions in the context of sliding mode control, providing another comprehensive reference for and discussion on generalized gradients and generalized solution concepts.

There are four definitions of the Dini derivative for functions $\varphi : \mathbb{R}^n \to \mathbb{R} \cup \{\pm\infty\}$. The upper right, lower right, upper left, and lower left Dini derivative at x in direction $w \in \mathbb{R}^n$ are defined as:

$$D^{+}\varphi(x; w) = \limsup_{v \to w;\ t \searrow 0} \frac{1}{t}\left(\varphi(x + tv) - \varphi(x)\right), \tag{2.19a}$$

$$D_{+}\varphi(x; w) = \liminf_{v \to w;\ t \searrow 0} \frac{1}{t}\left(\varphi(x + tv) - \varphi(x)\right), \tag{2.19b}$$

$$D^{-}\varphi(x; w) = \limsup_{v \to w;\ t \nearrow 0} \frac{1}{t}\left(\varphi(x + tv) - \varphi(x)\right), \tag{2.19c}$$

$$D_{-}\varphi(x; w) = \liminf_{v \to w;\ t \nearrow 0} \frac{1}{t}\left(\varphi(x + tv) - \varphi(x)\right). \tag{2.19d}$$

For Lipschitz continuous functions $\varphi : \mathbb{R}^n \to \mathbb{R}$ the Dini derivatives are finite for all $x \in \mathbb{R}^n$ and $w \in \mathbb{R}^n$, and the expressions (2.19) can be simplified by removing "$v \to w$" from the limit. For instance, when φ is Lipschitz, the upper right Dini derivative (2.19a) simplifies to

$$D^+\varphi(x; w) = \limsup_{t \searrow 0} \tfrac{1}{t} \left(\varphi(x + tw) - \varphi(x) \right). \qquad (2.20)$$

The three remaining Dini derivatives can be written in the same way. If φ is differentiable in $x \in \mathbb{R}^n$, then all Dini derivatives coincide with the directional derivative, i.e.,

$$\langle \nabla \varphi(x), w \rangle = D^+\varphi(x; w) = D_+\varphi(x; w) = D^-\varphi(x; w) = D_-\varphi(x; w). \quad (2.21)$$

Moreover, from Rademacher's theorem, [29, Theorem 3.2], it follows that a Lipschitz continuous function is differentiable almost everywhere.

The four definitions of the Dini derivative can indeed lead to different values for Lipschitz continuous functions.

Example 2.19 ([11, Example 2.3]) Let $\varphi : (-1, 1) \to \mathbb{R}$ be defined as

$$\varphi(x) = \begin{cases} x^2 \sin\left(x^{-1}\right), & \text{for } x \in (-1, 0), \\ 0, & \text{for } x = 0, \\ 2x^2 \sin\left(x^{-1}\right), & \text{for } x \in (0, 1). \end{cases} \qquad (2.22)$$

For $x \in (-1, 0)$ the derivative of φ is given by

$$\tfrac{d}{dx}\varphi(x) = 2x \sin\left(x^{-1}\right) - \cos\left(x^{-1}\right)$$

which can be estimated by

$$\sup_{x \in (-1,0)} \left| \tfrac{d}{dx}\varphi(x) \right| \le \sup_{x \in (-1,0)} \left(\left| 2x \sin(x^{-1}) \right| + \left| \cos(x^{-1}) \right| \right) \le 3.$$

In the same way we obtain

$$\sup_{x \in (0,1)} \left| \tfrac{d}{dx}\varphi(x) \right| \le 6 \quad \text{and} \quad |\varphi(x) - \varphi(0)| \le \left| 2x^2 \sin(x^{-1}) \right| \le 2\left| x^2 \right| \le 2|x - 0|$$

for all $x \in (-1, 1)$. Thus, the function φ is Lipschitz continuous with Lipschitz constant $L = 6$. For the Dini derivatives at $x = 0$ in direction $w = 1$ we obtain

$$\begin{aligned} D^+\varphi(0, 1) &= 2, & D^-\varphi(0, 1) &= 1, \\ D_+\varphi(0, 1) &= -2, & D_-\varphi(0, 1) &= -1. \end{aligned}$$

The function φ and the directional derivative for $x \in [-0.2, 0.2] \backslash \{0\}$ in the direction $w = 1$ are visualized in Fig. 2.5.

For an absolutely continuous solution $\phi(\cdot; x_0) \in \mathcal{S}(x_0)$ of the differential inclusion (2.1) and a real valued function $\varphi : \mathbb{R}^n \to \mathbb{R}$, consider the Dini derivatives for

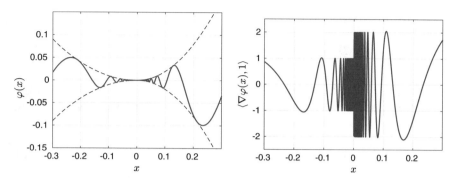

Fig. 2.5 Visualization of the function φ defined in (2.22) and its directional derivative on $[-0.3, 0.3]\backslash\{0\}$

$w = \dot{\phi}(t; x_0)$, at a fixed time t for which $\dot{\phi}(t; x_0)$ exists. Then the right Dini derivatives (2.19a)–(2.19b) indicates the change of φ in the direction of ϕ in the future time $t + \Delta t$, $\Delta t > 0$, whereas the left Dini derivatives (2.19c)–(2.19d) indicate the change for the past time $t - \Delta t$, $\Delta t > 0$. Consider, for instance, the simple example in which $n = 1$, $\phi(t; x_0) = t$ and $\varphi(x) = |x|$. Then the right Dini derivative at $t = 0$ equals 1, indicating that $t \mapsto \varphi(\phi(t; x_0))$ increases in t if future times $t > 0$ are considered, while the left Dini derivative at $t = 0$ equals -1, indicating that $t \mapsto \varphi(\phi(t; x_0))$ decreases with t for the past times $t < 0$.

Since stability and instability are defined by the future behavior of the solution, for characterizing stability properties of the differential inclusion (2.1) the right Dini derivatives are used. Nevertheless the left Dini derivative will be important in Sect. 4.3.

For a smooth function $\phi(\cdot; x_0) : \mathbb{R}_{\geq 0} \to \mathbb{R}^n$ and a smooth function $V : \mathbb{R}^n \to \mathbb{R}_{\geq 0}$ we use the notation

$$\dot{V}(\phi(t; x_0)) = \langle \nabla V(\phi(t; x_0)), \dot{\phi}(t; x_0) \rangle. \tag{2.23}$$

to indicate the derivative of V along the function ϕ. If ϕ is absolutely continuous and V is Lipschitz continuous, then (2.23) holds for almost all $t \in \mathbb{R}$.

Chapter 3
Strong (In)stability of Differential Inclusions and Lyapunov Characterizations

Abstract In this chapter we discuss extensions and generalizations of the stability and instability results of Sect. 2.2 for differential inclusions. In Sect. 3.1 $\mathcal{KL}$-stability and equivalent Lyapunov-characterizations are discussed. In Sect. 3.2 the results are mirrored to $\mathcal{K}_\infty \mathcal{K}_\infty$-instability. Connections between stability and instability are discussed in Sect. 3.3. In the last section of this chapter, Sect. 3.4, we briefly discuss $\mathcal{KL}$-stability with respect to two measures, which combines the concepts of stability and instability. The results discussed in this chapter correspond to *strong* stability notions, i.e., properties which are satisfied *for all* solutions of the differential inclusion (2.1). Thus, the results of this chapter describe robust stability properties of control systems as motivated for the dynamics (2.8). Weak stability results are discussed in Chap. 4.

Keywords Lyapunov methods · Differential inclusions · Stability of nonlinear systems · Instability of nonlinear systems · Stabilization/destabilization of nonlinear systems · Stabilizability and destabilizability

3.1 Strong $\mathcal{KL}$-Stability and Lyapunov Functions

In this section we discuss asymptotic stability of the origin of differential inclusions (2.1) as well as equivalent representations through $\mathcal{KL}$-stability and (robust) Lyapunov functions.

Definition 3.1 (*Global asymptotic stability*) The differential inclusion (2.1) is uniformly globally asymptotically stable with respect to the origin $0 \in \mathbb{R}^n$ if the following properties are satisfied. There exists a function $\delta \in \mathcal{K}_\infty$ such that for all $\varepsilon \geq 0$ and for all $\phi \in \mathcal{S}(x_0)$,

$$|\phi(t; x_0)| \leq \varepsilon \qquad \text{whenever } |x_0| \leq \delta(\varepsilon) \text{ and } t \geq 0,$$
$$|\phi(t; x_0)| \to 0 \qquad \text{for } t \to \infty.$$

© The Author(s), under exclusive license to Springer Nature Switzerland AG 2021
P. Braun et al., *(In-)Stability of Differential Inclusions*,
SpringerBriefs in Mathematics,
https://doi.org/10.1007/978-3-030-76317-6_3

Alternatively, uniform global asymptotic stability can be stated in terms of comparison functions known as (strong) $\mathcal{KL}$-stability.

Definition 3.2 (*Strong $\mathcal{KL}$-stability*) The differential inclusion (2.1) is *strongly $\mathcal{KL}$-stable with respect to the equilibrium* $0 \in \mathbb{R}^n$ if there exists $\beta \in \mathcal{KL}$, such that for all $x_0 \in \mathbb{R}^n$ every solution $\phi \in \mathcal{S}(x_0)$ satisfies

$$|\phi(t; x_0)| \leq \beta(|x_0|, t), \quad \forall\, t \in \mathbb{R}_{\geq 0}. \tag{3.1}$$

The term *strong* in Definition 3.2 refers to the fact that stability is required for all solutions and distinguishes this property from weak $\mathcal{KL}$-stability in Definition 4.2 in the next chapter. Whenever the difference is not important, *strong $\mathcal{KL}$-stability* is simply referred to as $\mathcal{KL}$-*stability*.

The two definitions, Definitions 3.1 and 3.2, are indeed equivalent, guaranteed by the following result [47, Proposition 2.5].

Theorem 3.3 *Consider the differential inclusion (2.1) satisfying Assumption 2.1. The differential inclusion is uniformly globally asymptotically stable with respect to the origin according to Definition 3.1 if and only if it is forward complete and is (strongly) $\mathcal{KL}$-stable according to Definition 3.2.*

The original statement in [47] is more general in terms of a forward invariant set instead of the origin. As pointed out in [47], the assumption that solutions of (2.1) are forward complete is redundant here due to the bound (3.1) in the context of stability of the origin. Note that in [47] systems of the form (2.8) with Lipschitz continuous right-hand side are discussed, and thus, [47, Proposition 2.5] is stated in terms of (2.8) instead of (2.1) satisfying Assumption 2.1. However, the proof of [47, Proposition 2.5] does not rely on the Lipschitz continuity and is also valid under the assumptions of Theorem 3.3.

As in Sect. 2.2 in the context of differential equations, asymptotic stability can be expressed in terms of appropriate Lyapunov functions.

Definition 3.4 (*(Robust) Lyapunov function*) A continuous function $V : \mathbb{R}^n \to \mathbb{R}$ is called a (robust) Lyapunov function for the differential inclusion (2.1) if there exist $\alpha_1, \alpha_2 \in \mathcal{K}_\infty$ and $\rho \in \mathcal{P}$ such that

$$\alpha_1(|x|) \leq V(x) \leq \alpha_2(|x|) \tag{3.2}$$

$$\max_{w \in F(x)} D^+ V(x; w) \leq -\rho(|x|) \tag{3.3}$$

holds for all $x \in \mathbb{R}^n$.

Similar to *strong* and *weak $\mathcal{KL}$-stability*, the word *robust* is used in some works to distinguish between *Lyapunov functions* and *control Lyapunov functions* (see Definition 4.4). While we have defined the decrease condition (3.3) in terms of the upper right Dini derivative, a specification of the Dini derivative is not necessary if V is smooth as pointed out in (2.21). The assumption on the existence of a smooth Lyapunov function in addition to continuity properties is not restrictive.

Theorem 3.5 (Lyapunov and converse Lyapunov theorem) *Consider the differential inclusion (2.1) and suppose that F satisfies Assumption 2.1. Then the following are equivalent.*

- *The differential inclusion is strongly $\mathcal{KL}$-stable according to Definition 3.2 with respect to the origin.*
- *There exists a smooth Lyapunov function according to Definition 3.4.*

The result in more general forms than presented here can be found in [47, Theorem 2.9], [23], [69, Theorem 1] or [20, Theorem. 1.1], for example. The result shows that for strong $\mathcal{KL}$-stability it is possible to assume that the robust Lyapunov function is smooth. Theorem 3.5 extends the classical stability result for ordinary differential equations, i.e., Theorem 2.10 and its converse statement.

3.2 $\mathcal{K}_\infty\mathcal{K}_\infty$-Instability and Chetaev Functions

In this section we mirror existing stability results from the preceding section to describe *strong complete instability* properties of equilibria of differential inclusions. To this end, we use definitions similar to Definitions 3.1 and 3.2 by using appropriate comparison functions and we give results analogous to Theorems 3.3 and 3.5.

With respect to instability, we are interested in the case where all solutions that do not start at the origin go to infinity. Moreover, solutions starting far away from the origin are not allowed to get arbitrarily close to the origin.

Definition 3.6 (*Strong complete instability*) The differential inclusion (2.1) is strongly completely unstable with respect to the origin $0 \in \mathbb{R}^n$ if the following properties are satisfied. There exists a function $\delta \in \mathcal{K}_\infty$ such that for all $\varepsilon > 0$ and for all solutions $\phi \in \mathcal{S}(x_0)$,

$$|\phi(t; x_0)| \geq \varepsilon \qquad \text{for all } t \geq 0, \qquad (3.4)$$
$$|\phi(t; x_0)| \to \infty \qquad \text{for } t \to \infty, \qquad (3.5)$$

whenever $|x_0| \geq \delta(\varepsilon)$.

In Definition 3.6 attention is not restricted to solutions which exist for all $t \geq 0$. If $\lim_{t \nearrow t^*} |\phi(t; x_0)| = \infty$ is satisfied for $t^* < \infty$, then we also consider (3.5) satisfied. This is consistent with the definition of solutions in Sect. 2.1 as extended real valued functions.

In the context of asymptotic stability, $\mathcal{KL}$-functions provide an upper bound for solutions of differential inclusions. To establish instability, by contrast, a lower bound for the solutions is needed.

Definition 3.7 ($\mathcal{K}_\infty\mathcal{K}$- *and* $\mathcal{K}_\infty\mathcal{K}_\infty$-*functions*) Consider the continuous function $\kappa : \mathbb{R}^2_{\geq 0} \to \mathbb{R}_{\geq 0}$.

- The function κ is said to be of class $\mathcal{K}_\infty\mathcal{K}$ ($\kappa \in \mathcal{K}_\infty\mathcal{K}$) if $\kappa(\cdot, s) \in \mathcal{K}_\infty$ for all $s \in \mathbb{R}_{\geq 0}$ and $\kappa(s, \cdot) - \kappa(s, 0) \in \mathcal{K}$ for all $s \in \mathbb{R}_{>0}$.
- The function κ is said to be of class $\mathcal{K}_\infty\mathcal{K}_\infty$ ($\kappa \in \mathcal{K}_\infty\mathcal{K}_\infty$) if $\kappa(\cdot, s) \in \mathcal{K}_\infty$ for all $s \in \mathbb{R}_{\geq 0}$ and $\kappa(s, \cdot) - \kappa(s, 0) \in \mathcal{K}_\infty$ for all $s \in \mathbb{R}_{>0}$.

Example 3.8 As an example, for $c > 0$ and $\lambda > 0$, consider the function $h : \mathbb{R}_{\geq 0}^2 \to \mathbb{R}_{\geq 0}$,

$$h(s, t) = ce^{\lambda t}s.$$

It holds that $h(\cdot, t) \in \mathcal{K}_\infty$ for all $t \in \mathbb{R}_{\geq 0}$. Additionally, $t \mapsto h(s, t) - h(s, 0) = c(e^{\lambda t} - 1)s$ is a $\mathcal{K}_\infty$ function for all $s > 0$ and $\lambda > 0$, and $h(s, \cdot) \in \mathcal{L}$ for all $s \geq 0$ and $\lambda < 0$. Thus $h \in \mathcal{K}_\infty\mathcal{K}_\infty$ for $\lambda > 0$ and $h \in \mathcal{K}\mathcal{L}$ for $\lambda < 0$.

With these definitions, we can define a counterpart to Definition 3.2 to characterize complete instability of the origin for differential inclusions.

Definition 3.9 (*Strong* $\mathcal{K}_\infty\mathcal{K}_\infty$-*instability*) Differential inclusion (2.1) is strongly $\mathcal{K}_\infty\mathcal{K}_\infty$-unstable with respect to the origin $0 \in \mathbb{R}^n$ if there exists $\kappa \in \mathcal{K}_\infty\mathcal{K}_\infty$ such that, for all $x_0 \in \mathbb{R}^n$ every solution $\phi \in \mathcal{S}(x_0)$ satisfies

$$|\phi(t; x_0)| \geq \kappa(|x_0|, t), \quad \forall\, t \in \mathbb{R}_{\geq 0}. \tag{3.6}$$

Remark 3.10 Note that in the literature the term $\mathcal{K}\mathcal{L}$-stability has been established, despite the fact that, for uniform global asymptotic stability, $\beta(\cdot, t)$ generally needs to be of class $\mathcal{K}_\infty$ (not only class $\mathcal{K}$) for all $t \geq 0$.

In Definition 3.9 one might wonder if $\kappa \in \mathcal{K}_\infty\mathcal{K}_\infty$ can be relaxed to $\kappa \in \mathcal{K}_\infty\mathcal{K}$. The following example illustrates why $\kappa \in \mathcal{K}_\infty\mathcal{K}$ does not provide an appropriate lower bound to ensure instability.

Example 3.11 Consider the ordinary differential equation $\dot{x} = 0$ which trivially has the origin as a stable equilibrium. Assume that $\kappa \in \mathcal{K}_\infty\mathcal{K}$ is used in Definition 3.9 to define complete instability and consider the function

$$\kappa(r, t) = \tfrac{1}{2}r(2 - e^{-t}) \in \mathcal{K}_\infty\mathcal{K} \setminus \mathcal{K}_\infty\mathcal{K}_\infty.$$

For all $x_0 \in \mathbb{R}^n$ and for all $t \in \mathbb{R}_{\geq 0}$ it holds that

$$|\phi(t; x_0)| = |x_0| \geq \tfrac{1}{2}|x_0|(2 - e^{-t}) = \kappa(|x_0|, t)$$

which would imply that the origin is unstable if we allowed for $\kappa \in \mathcal{K}_\infty\mathcal{K}$ in Definition 3.9. Since, however, the origin of the ODE is stable (though not asymptotically or $\mathcal{K}\mathcal{L}$-stable), $\mathcal{K}_\infty\mathcal{K}$-functions are not the right conceptual tool to describe equivalences between $\mathcal{K}\mathcal{L}$-stability and complete instability.

If only local instability is of interest, Definition 3.9 can be adapted as follows, for example.

Definition 3.12 Let $O \subset \mathbb{R}^n$ be an open neighborhood containing the origin $0 \in O$. The equilibrium $0 \in \mathbb{R}^n$ is locally strongly completely unstable with respect to the differential inclusion (2.1) and the neighborhood O if there exists a $\kappa \in \mathcal{K}_\infty \mathcal{K}_\infty$ such that, for all $x_0 \in O$ every solution $\phi \in \mathcal{S}(x_0)$ satisfies

$$|\phi(t; x_0)| \geq \kappa(|x_0|, t), \tag{3.7}$$

for all $t \in \mathbb{R}_{\geq 0}$ such that $\phi(t; x_0) \in O$.

Similar to Theorem 3.3 in the context of stability, a result establishing equivalence between Definitions 3.6 and 3.9 can be derived.

Theorem 3.13 *Consider the differential inclusion (2.1) satisfying Assumption 2.1. The origin is strongly completely unstable according to Definition 3.6 if and only if the origin is strongly $\mathcal{K}_\infty \mathcal{K}_\infty$-unstable according to Definition 3.9*

The result and its proof follow the lines inspired by [47, Proposition 2.5] which is given as Theorem 3.3 here. The proof is given in Sect. 6.1.

As in the preceding section, strong complete instability, or equivalently, strong $\mathcal{K}_\infty \mathcal{K}_\infty$-instability can be characterized through (strong) Chetaev functions.

Definition 3.14 (*(Robust) Chetaev function*) A continuous function $C : \mathbb{R}^n \to \mathbb{R}$ is called a Chetaev function for the differential inclusion (2.1) if there exist $\alpha_1, \alpha_2 \in \mathcal{K}_\infty$ and $\rho \in \mathcal{P}$ such that

$$\alpha_1(|x|) \leq C(x) \leq \alpha_2(|x|), \tag{3.8}$$

$$\min_{w \in F(x)} D_+ C(x; w) \geq \rho(|x|), \tag{3.9}$$

holds for all $x \in \mathbb{R}^n$.

A definition similar to the function C in Definition 3.14 and its existence in terms of a continuous and locally Lipschitz functions corresponding to a control system (2.8) has been discussed in [28, Lemma 1] and has been used in [26, 27]. As discussed after Example 2.16, Lyapunov also studied the type of function described in Definition 3.14. Hence, the function C in Definition 3.14 could also be called a Lyapunov function for instability. However, as motivated in the introduction and after Example 2.16 we use the term Lyapunov function in the context of stability and the term Chetaev function in the context of instability and in particular deviate from the terminology in [26–28].

Theorem 3.15 *Consider the differential inclusion (2.1) and suppose that F satisfies Assumption 2.1. Then the following are equivalent.*

- *The differential inclusion (2.1) is strongly $\mathcal{K}_\infty \mathcal{K}_\infty$-unstable according to Definition 3.9.*
- *There exists a smooth Chetaev function according to Definition 3.14.*

The proof of Theorem 3.15 follows the lines of [69] showing a generalization of Theorem 3.5 by reversing the inequalities. The complete proof can be found in Sect. 6.2, here we only comment on some of its aspects.

Theorem 3.15 is shown by proving a stronger result. The right-hand side of (2.1) is embedded in a larger set $F(x) \subset F_L(x)$ for all $x \in \mathbb{R}^n$ where $F_L : \mathbb{R}^n \rightrightarrows \mathbb{R}^n$ satisfies Assumptions 2.1 and 2.2, i.e, F_L is in particular Lipschitz continuous. Then, the equivalence in Theorem 3.15 is shown for $\dot{x} \in F_L(x)$ from which the result for (2.1) is concluded.

To construct a Chetaev function, we use the inequalities

$$\alpha_2(|\phi(t; x_0)|) \geq \alpha_2(\kappa(|x_0|, t)) \geq \alpha_1(|x_0|)e^{2t},$$

which hold for all $x \in \mathbb{R}^n$ and all $\phi \in S(x_0)$, with $\alpha_1, \alpha_2 \in \mathcal{K}_\infty$ from Lemma 7.2 and $\kappa \in \mathcal{K}_\infty \mathcal{K}_\infty$ from Definition 3.9. Based on this estimate, it is shown that

$$C_1(x_0) = \inf_{t \geq 0; \ \phi \in S_L(x_0)} \alpha_2(|\phi(t; x_0)|)e^{-t} \tag{3.10}$$

is well-defined, is in fact continuous on $\mathbb{R}^n$ and locally Lipschitz continuous on $\mathbb{R}^n \backslash \{0\}$, and is a Chetaev function satisfying the properties of Definition 3.14. As a last step smoothing techniques are applied to obtain a smooth Chetaev function C from C_1.

3.3   Relations Between Chetaev Functions, Lyapunov Functions, and Scaling

In Theorem 2.3 we have discussed that a scaling of the differential inclusion (2.1) by means of a positive function $\eta : \mathbb{R}_{\geq 0} \to \mathbb{R}_{>0}$ preserves the stability properties of (2.1) with respect to the origin. In particular, the origin of $\dot{x} \in \eta(|x|)F(x)$ is strongly $\mathcal{K}\mathcal{L}$-stable (strongly $\mathcal{K}_\infty \mathcal{K}_\infty$-unstable) if and only if the origin of $\dot{x} \in \eta(|x|)F(x)$ is strongly $\mathcal{K}\mathcal{L}$-stable (strongly $\mathcal{K}_\infty \mathcal{K}_\infty$-unstable). This is immediately clear by considering the Lyapunov representation of strong $\mathcal{K}\mathcal{L}$-stability and Chetaev representation of strong $\mathcal{K}_\infty \mathcal{K}_\infty$-instability, respectively.

Lemma 3.16 *Consider the differential inclusion (2.1) satisfying Assumption 2.1 together with its scaled version (2.3) for a Lipschitz continuous positive scaling* $\eta : \mathbb{R}_{\geq 0} \to \mathbb{R}_{>0}$.

- *Assume that $V : \mathbb{R}^n \to \mathbb{R}_{\geq 0}$ is a smooth Lyapunov function for (2.1) according to Definition 3.4. Then V is a smooth Lyapunov function of the scaled differential inclusion (2.3).*
- *Assume that $C : \mathbb{R}^n \to \mathbb{R}_{\geq 0}$ is a smooth Chetaev function for (2.1) according to Definition 3.14. Then C is a smooth Chetaev function of the scaled differential inclusion (2.3).*

Proof Let V denote a smooth Lyapunov function for (2.1). Then there exists $\rho \in \mathcal{P}$ such that in particular the inequality

$$\max_{w \in F(x)} \langle \nabla V(x), w \rangle \leq -\rho(|x|) \tag{3.11}$$

is satisfied for all $x \in \mathbb{R}^n$. Thus, the inequality

$$\max_{w \in \eta(|x|)F(x)} \langle \nabla V(x), w \rangle = \max_{w \in F(x)} \langle \nabla V(x), \eta(|x|)w \rangle \leq -\eta(|x|)\rho(|x|)$$

holds for all $x \in \mathbb{R}^n$ and the function $\tilde{\rho}(|x|) = \eta(|x|)\rho(|x|)$ satisfies $\tilde{\rho} \in \mathcal{P}$. Hence V is a Lyapunov function for (2.3). The second assertion follows similarly. □

The particular scaling in Corollary 2.4 additionally ensures that all solutions of the differential inclusion (2.3) are forward complete. This fact is crucial for the proof of Theorem 3.15.

While the scaling of the differential inclusion leads to a scaling of the decrease of the Lyapunov or the Chetaev function, i.e., the right-hand side in (3.3) and (3.9), respectively, also the Lyapunov and the Chetaev function can be scaled to obtain a certain decrease. We will now use this fact in order to show that, after appropriate rescaling, (3.9) can be written as $\min_{w \in F(x)} D_+\widehat{C}(x; w) \geq \widehat{C}(x)$.

To this end, observe that for $\hat{\rho} = \rho \circ \alpha_2^{-1} \in \mathcal{P}$, using the upper bound in (3.8), it holds that

$$\min_{w \in F(x)} D_+C(x; w) \geq \rho(|x|) \geq \rho(\alpha_2^{-1}(C(x))) = \hat{\rho}(C(x)).$$

Furthermore, if we select an $\hat{\alpha} \in \mathcal{K}_\infty$ continuously differentiable such that $\hat{\alpha}'(s) > 0$ for all $s \in \mathbb{R}_{>0}$ and

$$\hat{\alpha}(s) \leq \hat{\rho}(s)\hat{\alpha}'(s) \quad \forall s \in \mathbb{R}_{>0},$$

an additional increase condition can be derived based on the function $\widehat{C}(x) = \hat{\alpha}(C(x))$. An $\hat{\alpha}$ satisfying these properties is guaranteed to exist according to Lemma 7.6, [39, Lemma 18]. First note that from the results derived in [62], the applicability of the chain rule with respect to the Dini derivative follows, and thus

$$D_+\widehat{C}(x; w) = \hat{\alpha}'(C(x))D_+C(x; w) \quad \forall w \in \mathbb{R}^n. \tag{3.12}$$

Using (3.12) the chain of inequalities

$$\min_{w \in F(x)} D_+\widehat{C}(x; w) \geq \hat{\alpha}'(C(x))\hat{\rho}(C(x)) \geq \hat{\alpha}(C(x)) = \widehat{C}(x) \tag{3.13}$$

is satisfied which shows that by appropriately selecting the Chetaev function, the function ρ characterizing the increase in (3.9) can be replaced by the Chetaev function itself. To show that $\widehat{C}$ is indeed a Chetaev function, we additionally consider $\hat{\alpha}_1 = \hat{\alpha} \circ \alpha_1$, $\hat{\alpha}_2 = \hat{\alpha} \circ \alpha_2$, satisfying $\hat{\alpha}_1, \hat{\alpha}_2 \in \mathcal{K}_\infty$ and

$$\hat{\alpha}_1(|x|) \leq \widehat{C}(x) \leq \hat{\alpha}_2(|x|) \qquad \forall\, x \in \mathbb{R}^n,$$

showing that $\widehat{C}$ can be lower and upper bounded by $\mathcal{K}_\infty$ functions, ensuring the conditions (3.8). Moreover, if (3.13) is satisfied, then (3.9) can be obtained through $\rho(s) := \min_{s=|x|} \widehat{C}(x)$.

While we have shown these derivations in terms of Chetaev functions, the same reasoning can be applied for Lyapunov functions. This scaling of Lyapunov functions is also described in [36, Sect. 2.1], for example. We summarize the facts discussed here in the following remark. The scaling enables us to use Lyapunov and Chetaev functions satisfying specific decrease/increase conditions without loss of generality in the proofs of the main results in Chap. 6.

Remark 3.17 If C is a Chetaev function, then any nonlinear scaling $\widehat{C}(x) = \alpha(C(x))$, $\alpha \in \mathcal{K}_\infty$, provides a Chetaev function. Moreover, through an appropriate scaling $\alpha \in \mathcal{K}_\infty$ condition (3.9) can be replaced by

$$\min_{w \in F(x)} D_+\widehat{C}(x; w) \geq \widehat{C}(x). \tag{3.14}$$

In the same way, if V is a Lyapunov function, (3.3) can be replaced by

$$\max_{w \in F(x)} D^+\widehat{V}(x; w) \leq -\widehat{V}(x). \tag{3.15}$$

The bounds (3.2) and (3.8) are satisfied for $\tilde{\alpha}_1 = \alpha \circ \alpha_1 \in \mathcal{K}_\infty$ and $\tilde{\alpha}_2 = \alpha \circ \alpha_2 \in \mathcal{K}_\infty$, and (3.9) and (3.3) can be obtained using $\rho(s) = \min_{s=|x|} \widehat{C}(x)$ and $\rho(s) = \min_{s=|x|} \widehat{V}(x)$ from (3.14) and (3.15), respectively.

A second property that follows immediately from the Lyapunov and the Chetaev characterizations relates (2.1) and its time reversed system (2.2). In particular, if V is a smooth Lyapunov function for $\dot{x} \in F(x)$, then $C = V$ is a smooth Chetaev function for $\dot{x} \in -F(x)$ and vice versa.

Corollary 3.18 *Consider the differential inclusion (2.1) satisfying Assumption 2.1 together with its time reversed counterpart (2.2).*

- *Assume that $V : \mathbb{R}^n \to \mathbb{R}_{\geq 0}$ is a smooth Lyapunov function for (2.1) according to Definition 3.4. Then $C = V$ is a smooth Chetaev function for (2.2).*
- *Assume that $C : \mathbb{R}^n \to \mathbb{R}_{\geq 0}$ is a smooth Chetaev function for (2.1) according to Definition 3.14. Then $V = C$ is a smooth Lyapunov function for (2.2).*

Proof Let V denote a smooth Lyapunov function for (2.1). Then there exists $\rho \in \mathcal{P}$ such that the inequality

$$-\rho(|x|) \geq \max_{w \in F(x)} \langle \nabla V(x), w \rangle = -\min_{w \in F(x)} -\langle \nabla V(x), w \rangle \tag{3.16}$$

is satisfied for all $x \in \mathbb{R}^n$. Multiplying both sides by -1 allows us to rewrite the inequality as

$$\rho(|x|) \geq \min_{w \in F(x)} -\langle \nabla V(x), w \rangle = \min_{w \in F(x)} \langle \nabla V(x), -w \rangle = \min_{w \in -F(x)} \langle \nabla V(x), w \rangle$$

which shows that $C = V$ is a Chetaev function for (2.2). The second item follows the same lines.   □

While connections between Chetaev functions and Lyapunov functions presented in this section are rather simple and illustrative, connections between the stability notions using $\mathcal{KL}$-functions and $\mathcal{K}_\infty \mathcal{K}_\infty$-functions are in general not straightforward. In particular, it may be nontrivial to construct a $\mathcal{K}_\infty \mathcal{K}_\infty$-function for strong complete instability of $\dot{x} \in -F(x)$ from a $\mathcal{KL}$-function for $\mathcal{KL}$-stability of $\dot{x} \in F(x)$. This is because the solutions of $\dot{x} \in F(x)$ in backward and in forward time may differ significantly when F is nonlinear. We illustrate this fact by a simple example.

Example 3.19 Consider the differential equation $\dot{x} = -x^3$ as a special case of the differential inclusion (2.1). The differential equation represents the time reversal system of the dynamics discussed in Example 2.5. The solution of the differential equation is given by

$$\phi(t; x_0) = -\frac{x_0}{\sqrt{1 - 2tx_0^2}}.$$

Thus, using the $\mathcal{KL}$-function

$$\beta(r, t) = \frac{r}{\sqrt{1 - 2tr^2}}$$

the differential equation $\dot{x} = -x^3$ is (strongly) $\mathcal{KL}$-stable according to Definition 3.2. Alternatively, we can define the Lyapunov function $V(x) = \frac{1}{2}x^2$ which satisfies

$$\alpha_1(|x|) = \tfrac{1}{2}|x|^2 \leq V(x) \leq \tfrac{1}{2}|x|^2 = \alpha_2(|x|),$$
$$\langle \nabla V(x), -x^3 \rangle \leq -|x|^4 = -\rho(|x|),$$

and thus $\mathcal{KL}$-stability can be concluded from Theorem 3.5.

From Corollary 3.18 it follows that $C = V$ is a Chetaev function with respect to the time reversal dynamics $\dot{x} = x^3$ and (strong) $\mathcal{K}_\infty \mathcal{K}_\infty$-stability follows from Theorem 3.15. The solution of $\dot{x} = x^3$ is given by

$$\psi(t; x_0) = \frac{x_0}{\sqrt{1 - 2tx_0^2}} \tag{3.17}$$

as discussed in Example 2.5 and since the solutions admit finite escape time (3.17) cannot be used directly to define a $\mathcal{K}_\infty\mathcal{K}_\infty$-function such that (3.6) is satisfied.

However, a bound of the form $|\phi(t; x_0)| \geq \kappa(|x_0|, t)$, for $\kappa \in \mathcal{K}_\infty\mathcal{K}_\infty$ is still possible for all $t \in \mathbb{R}_{\geq 0}$ where $\kappa(|x_0|, t)$ is finite for all $t \in \mathbb{R}_{\geq 0}$ and for all $x_0 \in \mathbb{R}$ and possibly $|\phi(t; x_0)| = \infty$ depending on t.

Note that finite escape time does not create a problem in (3.9), i.e., the left-hand side and the right-hand side of

$$\langle \nabla C(x), x^3 \rangle \geq |x|^4 = \rho(|x|),$$

are finite for all $x \in \mathbb{R}$ and for $C(x) = \frac{1}{2}x^2$. Nevertheless, a $\mathcal{K}_\infty\mathcal{K}_\infty$-function providing a lower bound for the solutions exists according to Theorem 3.15 and it can be shown that

$$|\psi(t; x_0)| \geq \kappa(|x_0|, t), \qquad \forall t \geq 0 \tag{3.18}$$

and where $\kappa \in \mathcal{K}_\infty\mathcal{K}_\infty$ is defined as

$$\kappa(|x_0|, t) \geq \begin{cases} |x_0|\sqrt{t+1} & \text{if } |x_0| \geq 1, \\ |x_0|^2\sqrt{t+1} & \text{if } |x_0| \leq 1. \end{cases} \tag{3.19}$$

Claim 3.20 *The solution of the differential equation $\dot{x} = x^3$ satisfies (3.18) for κ defined in (3.19) and this κ is a $\mathcal{K}_\infty\mathcal{K}_\infty$ function.*

Proof of Claim 3.20 For $x_0 = 0$, the inequality $|\psi(t; x_0)| \geq \kappa(|x_0|, t)$ is trivially satisfied for all $t \in \mathbb{R}_{\geq 0}$. For $|x_0| \geq 1$ and $0 \leq t < \frac{1}{2x_0^2}$, it holds that

$$1 \geq 1 + (1 - 2x_0^2)t - 2x_0^2t^2 = (t+1)(1 - 2x_0^2t)$$

which implies the inequality

$$\frac{x_0^2}{1 - 2x_0^2t} \geq x_0^2(t+1).$$

Taking the square root on both sides, we can conclude that $|\psi(t; x_0)| \geq |x_0|\sqrt{t+1}$ for all $|x_0| \geq 1$ and for all $t \geq 0$.

To show that $|\psi(t; x_0)| \geq \kappa(|x_0|, t)$ for $|x_0| \leq 1$, first note that

$$-7 - 4|x_0| + 8x_0^2 < 0$$

for all $0 < |x_0| \leq 1$. Thus, it can be shown that the polynomial

$$p(t) = -2x_0^2t^2 + (1 - 2x_0)t - \frac{1 - x_0^2}{x_0^2}$$

satisfies $p(t) \neq 0$ for all $t \in \mathbb{R}$. Moreover, from $\lim_{t \to \infty} p(t) = -\infty$, it follows that $p(t) \leq 0$ for all $t \geq 0$. Hence, for all $t \geq 0$ it holds that

$$\frac{1 - x_0^2}{x_0^2} = -1 + \frac{1}{x_0^2} \geq -2x_0^2 t^2 + (1 - 2x_0)t = (1 - 2x_0^2)t - 2x_0^2 t^2$$

and rearranging the terms leads to the expression

$$1 \geq x_0^2(t + 1 - 2x_0^2 t^2 - 2x_0^2 t) = (1 - 2x_0^2 t)x_0^2(t + 1).$$

Finally, for $0 \leq t < \frac{1}{2x_0^2}$ the inequality

$$\frac{x_0^2}{1 - 2x_0^2 t} \geq x_0^4(t + 1)$$

is satisfied and taking the square root on both sides shows (3.18). The fact that $\kappa \in \mathcal{K}_\infty \mathcal{K}_\infty$ follows immediately from (3.19) and Definition 3.7. □

3.4 $\mathcal{K}\mathcal{L}$-Stability with Respect to (Two) Measures

While we restrict our presentation to stability of the origin with respect to the Euclidean norm, less restrictive generalizations of $\mathcal{K}\mathcal{L}$-stability and corresponding results extending Theorem 3.5 can be obtained. In particular, the Euclidean norm can be replaced by a so-called measure, i.e., a positive function, and the origin can be replaced by a forward invariant set. Here, we briefly discuss the main result of [69] in the framework of this monograph, as this appears to be the most general result available. Nevertheless, as also pointed out in [69], earlier ideas date back to [45, 46, 55]. Results on $\mathcal{K}\mathcal{L}$-stability with respect to two measures [69, Theorem 1], combine stability and instability properties in a single framework without the distinction between Lyapunov and Chetaev functions and without considering $\mathcal{K}_\infty \mathcal{K}_\infty$-functions.

The two measures $\omega_1, \omega_2 : \mathcal{G} \to \mathbb{R}_{\geq 0}$ are positive functions from an open set $\mathcal{G} \subset \mathbb{R}^n$ to the positive real numbers. Then (2.1) is called $\mathcal{K}\mathcal{L}$-stable with respect to (ω_1, ω_2) on $\mathcal{G}$ if there exists a $\mathcal{K}\mathcal{L}$-function β such that for all $x \in \mathcal{G}$, (see [69, Definition 6])

$$\begin{aligned} \omega_1(\phi(t; x_0)) &\leq \beta(\omega_2(x_0), t) \qquad \forall\, t \geq 0 \\ \text{and} \qquad \phi(t; x_0) &\in \mathcal{G} \qquad \forall \phi \in \mathcal{S}(x_0) \quad \forall\, t \geq 0. \end{aligned} \tag{3.20}$$

For $\mathcal{G} = \mathbb{R}^n$ and $\omega_1(x) = \omega_2(x) = |x|$ the Definition 3.2 for $\mathcal{K}\mathcal{L}$-stability of the origin is recovered. Moreover, if we consider an open domain $\mathcal{G} \subset \mathbb{R}^n \backslash \{0\}$ excluding the origin, then measures of the form $\omega_1(x) = \omega_2(x) = \frac{1}{|x|}$ can be considered to describe certain instability properties. For the second selection of measures the inequality (3.20) can be rewritten as

$$|\phi(t; x_0)| \geq \left(\beta \left(\left| \tfrac{1}{x_0} \right|, t \right) \right)^{-1}$$

which provides a lower bound on ϕ similar to (3.7).

For a Lyapunov function characterizing $\mathcal{KL}$-stability with respect to (ω_1, ω_2), the condition (3.2) needs to be replaced by

$$\alpha_1(\omega_1(x)) \leq V(x) \leq \alpha_2(\omega_2(x)).$$

With this modification, an extension of Theorem 3.5 to $\mathcal{KL}$-stability with respect to two measures can be obtained, see [69, Theorem 1].

While [69, Theorem 1] and related results with respect to two measures are more versatile than the results discussed here, we believe that the consideration of Chetaev functions adds another perspective to these results which is significantly more intuitive in the context of instability when solutions diverge from the equilibrium. In particular, $\mathcal{KL}$-stability with respect to $|\cdot|^{-1}$ focuses on "stability of $x = \infty$" without acknowledging the special role of the origin $x = 0$. In contrast, $\mathcal{K}_\infty \mathcal{K}_\infty$-instability focuses on "instability around $x = 0$" where states far from the origin are not important. This is also important for the weak stability/instability setting discussed in the next chapter, where weak stability with respect to two measures has not been discussed to the best of our knowledge.

We illustrate some equivalences of and differences between strong $\mathcal{K}_\infty \mathcal{K}_\infty$-instability and $\mathcal{KL}$-stability with respect to $\omega_1(x) = \omega_2(x) = \frac{1}{|x|}$ in the following based on simple examples.

Example 3.21 Consider the differential equation $\dot{x} = x$ as a special form of the differential inclusion (2.1) with unique solution $\phi(t; x_0) = e^t x_0$.

The origin is strongly $\mathcal{K}_\infty \mathcal{K}_\infty$-unstable according to Definition 3.9 and with $\mathcal{K}_\infty \mathcal{K}_\infty$-function $\kappa(r, t) = e^t r$. Moreover, since $C(x) = \alpha_1(|x|) = \alpha_2(|x|) = \frac{1}{2}x^2$ and

$$\langle \nabla C(x), \dot{x} \rangle = x^2 \geq C(x)$$

are satisfied for all $x \in \mathbb{R}$ and C is a smooth strong Chetaev function according to Definition 3.14.

Similarly, if we define the $\mathcal{KL}$-function $\beta(s, t) = e^{-t}s$ and if we exclude the origin, i.e., $\mathcal{G} = \mathbb{R} \backslash \{0\}$, then it holds that

$$\omega_1(|\phi(t; x_0)|) = \tfrac{1}{|x_0|}e^{-t} = \beta\left(\tfrac{1}{|x_0|}, t \right) = \beta\left(\omega_2(|x_0|), t \right)$$

which shows $\mathcal{KL}$-stability with respect to the measure $|\cdot|^{-1}$. Moreover, the lower bounds from $\mathcal{KL}$-stability and $\mathcal{K}_\infty\mathcal{K}_\infty$-instability given by $|\phi(t; x_0)| \geq e^t |x_0|$ coincide.

To obtain a Lyapunov function we define $V(x) = \alpha_1(\omega_1(x)) = \alpha_2(\omega_2(x)) = \frac{1}{2}x^{-2}$ for $x \neq 0$. Then it holds that

$$\langle \nabla V(x), \dot{x} \rangle = -x^{-2} \leq -V(x)$$

and which shows that V is a Lyapunov function with respect to the measure $|\cdot|^{-1}$.

In this example, the only differences between $\mathcal{KL}$-stability with respect to $|\cdot|^{-1}$ and $\mathcal{K}_\infty\mathcal{K}_\infty$-instability are that the origin needs to be excluded from the consideration in the context of $\mathcal{KL}$-stability and that the rate of change of the Lyapunov function and the Chetaev function along solutions differ. More precisely, $\dot{V}(\phi(t; x_0))$ is large for $\phi(t; x_0)$ close to the origin and small for states far from the origin, while $\dot{C}(\phi(t; x_0))$ behaves exactly in the opposite way.

Example 3.22 Consider the differential equation $\dot{x} = x^3$ with solution

$$\phi(t; x) = \frac{x}{\sqrt{1 - 2tx^2}},$$

which was already discussed in detail in Example 3.19. Even though solutions exhibit finite escape time, a bound of the form $|\phi(t; x_0)| \geq \kappa(t, |x_0|)$ for $\kappa \in \mathcal{K}_\infty\mathcal{K}_\infty$ can be established for all $t \in \mathbb{R}_{\geq 0}$ if we set $|\phi(t; x_0)| := \infty$ for t beyond the finite escape time, as introduced in Sect. 2.1. Independent of this extension, the condition (3.9), i.e.,

$$\langle \nabla C(x), x^3 \rangle \geq \rho(|x|),$$

is well defined for all $x \in \mathbb{R}$ and is in particular satisfied for $C(x) = x^2$ and $\rho(s) = 3s^4$, which is a $\mathcal{K}_\infty$ function.

In contrast, differential inclusions admitting finite escape time are not covered by [69, Theorem 1]. However, similar extensions of $\mathcal{KL}$-stability with respect to the measures $(|\cdot|^{-1}, |\cdot|^{-1})$ and finite escape time seem to be straightforward. In this case the instability property (3.6) would be captured through the bound

$$\frac{1}{|\phi(t; x_0)|} \leq \beta\left(|x_0|^{-1}, t\right)$$

where $|\phi(t; x_0)|^{-1} = 0$ in the case that $|\phi(t; x_0)| = \infty$.

Chapter 4
Weak (In)stability of Differential Inclusions and Lyapunov Characterizations

Abstract In the preceding chapter we have discussed *strong stability* and *strong instability* results for differential inclusions where the term *strong* refers to *all solutions* of a differential inclusion. In this chapter we discuss *weak stability* and *weak instability* results, where properties need to be satisfied for at least *one solution* instead of for *all solutions*. While the results from the last chapter allowed us to draw conclusions in terms of robustness, the results in this chapter guarantee stabilizability or destabilizability of the origin. In particular, in the context of a control system (2.8), weak stability guarantees the existence of an input $u(\cdot)$ such that the origin can be reached asymptotically and weak instability describes the fact that it is possible to steer away from the origin by an appropriate selection of $u(\cdot)$. In contrast to the last chapter where it turned out that strong stability in forward time is equivalent to strong instability in backward time, the same equivalence is not true for weak stability/instability results.

Keywords Lyapunov methods · Differential inclusions · Stability of nonlinear systems · Instability of nonlinear systems · Stabilization/destabilization of nonlinear systems · Stabilizability and destabilizability

4.1 Weak $\mathcal{KL}$-Stability and Control Lyapunov Functions

By replacing *for all* with *there exists* in Definitions 3.1 and 3.9 the corresponding weak notions of stability of the origin are obtained.

Definition 4.1 (*Global asymptotic stabilizability*) The differential inclusion (2.1) is uniformly globally asymptotically stabilizable with respect to the origin $0 \in \mathbb{R}^n$ if the following are satisfied. There exists a function $\delta \in \mathcal{K}_\infty$ such that for all $\varepsilon \geq 0$ and all $x_0 \in \mathbb{R}^n$ with $|x_0| \leq \delta(\varepsilon)$ there exists $\phi \in \mathcal{S}(x_0)$ with

$$|\phi(t; x_0)| \leq \varepsilon \qquad\qquad \text{for all } t \geq 0 \quad \text{and}$$
$$|\phi(t; x_0)| \to 0 \qquad\qquad \text{for } t \to \infty.$$

Definition 4.2 (*Weak $\mathcal{KL}$-stability*) The differential inclusion (2.1) is *weakly $\mathcal{KL}$-stable* with respect to the equilibrium $0 \in \mathbb{R}^n$ if there exists $\beta \in \mathcal{KL}$ such that, for all $x_0 \in \mathbb{R}^n$ there exists $\phi \in \mathcal{S}(x_0)$ with

$$|\phi(t; x_0)| \leq \beta(|x_0|, t), \quad \forall\, t \in \mathbb{R}_{\geq 0}. \tag{4.1}$$

Strong $\mathcal{KL}$-stability describes robustness properties of the equilibrium, whereas weak $\mathcal{KL}$-stability indicates that a system is stabilizable. By replacing *for all* with *there exists* an analogous result to Theorem 3.3 can be stated.

Corollary 4.3 *Consider the differential inclusion (2.1) satisfying Assumption 2.1. The differential inclusion is globally asymptotically stabilizable with respect to the origin according to Definition 4.1 if and only if it is weakly $\mathcal{KL}$-stable according to Definition 4.2.*

The proof of Corollary 4.3 follows from the proof of Theorem 3.3 in [47, Proposition 2.5] with minimal changes by replacing all solutions with particular solutions.

Definition 4.4 (*Control Lyapunov function*) A continuous function $V : \mathbb{R}^n \to \mathbb{R}$ is called a control Lyapunov function for the differential inclusion (2.1) if there exist $\alpha_1, \alpha_2 \in \mathcal{K}_\infty$ and $\rho \in \mathcal{P}$ such that

$$\alpha_1(|x|) \leq V(x) \leq \alpha_2(|x|) \tag{4.2}$$
$$\min_{w \in F(x)} D_+ V(x; w) \leq -\rho(|x|) \tag{4.3}$$

holds for all $x \in \mathbb{R}^n$.

In contrast to Definition 3.4, the upper right Dini derivative is replaced by the lower right Dini derivative here. Since V relies on the Dini derivative, in contrast to alternative definitions based, e.g., on the proximal gradient [20], V is also referred to *control Lyapunov function in the Dini sense* in the literature.

Theorem 4.5 *Suppose F satisfies Assumptions 2.1 and 2.2. Then the following are equivalent.*

- *The differential inclusion (2.1) is weakly $\mathcal{KL}$-stable according to Definition 4.2.*
- *There exists a Lipschitz continuous control Lyapunov function according to Definition 4.4.*

Among other papers, Theorem 4.5 can be found in different variants in [39, 59, 63, 66]. In [63] the original definition of a control Lyapunov function in the Dini sense along with initial results is given. In [66] the equivalence in Theorem 4.5 for a continuous control Lyapunov function is established before the existence of a

Lipschitz continuous function (in fact a semiconcave function) is shown in [59]. In [39], Theorem 4.5 is proven for a more general forward invariant set replacing the origin.

Theorems 3.5 and 4.5 extend the classical stability result for ordinary differential equations as given in Theorem 2.10. Since in the case of ordinary differential equations with Lipschitz continuous right-hand side $\mathcal{S}(x_0)$ contains only a single element, the definitions of strong and weak $\mathcal{KL}$-stability coincide and are equivalent to uniform global asymptotic stability [47, Proposition 2.5].

Note the difference in Theorems 3.5 and 4.5 in terms of differentiability of V. Indeed, in Theorem 4.5 it cannot be assumed that V is smooth. Classical examples clarifying this fact will be discussed in the next section.

4.2 Weak $\mathcal{K}_\infty\mathcal{K}_\infty$-Instability and Control Chetaev Functions

With the definitions of the preceding section, analogous definitions expressing weak instability properties are straightforward.

Definition 4.6 (*Weak complete instability*) The differential inclusion (2.1) is weakly completely unstable with respect to the origin $0 \in \mathbb{R}^n$ if the following properties are satisfied. There exists a function $\delta \in \mathcal{K}_\infty$ such that for all $\varepsilon > 0$ and all $x_0 \in \mathbb{R}^n$ with $|x_0| \geq \delta(\varepsilon)$ there exists $\phi \in \mathcal{S}(x_0)$ with

$$|\phi(t; x_0)| \geq \varepsilon \qquad \text{for all } t \geq 0 \quad \text{and} \qquad (4.4)$$
$$|\phi(t; x_0)| \to \infty \qquad \text{for } t \to \infty. \qquad (4.5)$$

Remark 4.7 Note that in Definition 4.6 condition (4.5) without (4.4) does not prevent solutions from passing through the origin. Indeed, the solutions $\phi(t; x_0) = x_0 + t$, $x_0 \in \mathbb{R}$, of the differential inclusion

$$\dot{x} \in \begin{cases} [0, 1] & \text{for } x = 0 \\ 1 & \text{for } x \neq 0 \end{cases}$$

satisfy (4.5) for all $x_0 \in \mathbb{R}$. The use of (4.4) instead of the weaker condition $|\phi(t; x_0)| \neq 0$ for all $t \geq 0$, for all $x_0 \neq 0$, is motivated by the discussion in Example 2.17.

Definition 4.8 (*Weak $\mathcal{K}_\infty\mathcal{K}_\infty$-instability*) The equilibrium $0 \in \mathbb{R}^n$ is weakly $\mathcal{K}_\infty\mathcal{K}_\infty$-unstable with respect to the differential inclusion (2.1) if there exists $\kappa \in \mathcal{K}_\infty\mathcal{K}_\infty$ such that, for all $x_0 \in \mathbb{R}^n$ there exists $\phi \in \mathcal{S}(x_0)$ so that

$$|\phi(t; x_0)| \geq \kappa(|x_0|, t) \quad \text{for all } t \geq 0. \qquad (4.6)$$

Corollary 4.9 *Consider the differential inclusion (2.1) satisfying Assumption 2.1. The differential inclusion is weakly completely unstable with respect to the origin according to Definition 4.6 if and only if it is is weakly $\mathcal{K}_\infty\mathcal{K}_\infty$-unstable according to Definition 4.8.*

The proof of Corollary 4.9 can be obtained from the proof of Theorem 3.13 given in Sect. 6.1 by replacing all solutions $\phi \in \mathcal{S}(x_0)$ with a particular solution $\phi \in \mathcal{S}(x_0)$.

Definition 4.10 (*Control Chetaev function*) A continuous function $C : \mathbb{R}^n \to \mathbb{R}$ is called a control Chetaev function for the differential inclusion (2.1) if there exist $\alpha_1, \alpha_2 \in \mathcal{K}_\infty$ and $\rho \in \mathcal{P}$ such that

$$\alpha_1(|x|) \le C(x) \le \alpha_2(|x|) \tag{4.7}$$

$$\max_{w \in F(x)} D^+C(x; w) \ge \rho(|x|) \tag{4.8}$$

holds for all $x \in \mathbb{R}^n$.

Similar to the equivalence between weak $\mathcal{KL}$-stability and the existence of control Lyapunov functions in Theorem 4.5, the main result of this Chapter shows the equivalence between weak complete instability and the existence of a continuous control Chetaev function.

Theorem 4.11 *Consider the differential inclusion (2.1) satisfying Assumptions 2.1 and 2.2. Then the following are equivalent.*

- *The origin of the differential inclusion (2.1) is weakly $\mathcal{K}_\infty\mathcal{K}_\infty$-unstable according to Definition 4.8.*
- *There exists a continuous control Chetaev function according to Definition 4.10.*

The proof of Theorem 4.11 is deferred to Sect. 6.3 and is based on the following ideas. The construction of the control Chetaev function uses the cost functional

$$J(x_0, \phi) = \begin{cases} \frac{1}{\int_0^\infty \frac{1}{g(\phi(t;x_0))}dt}, & \text{if } \int_0^\infty g(\phi(t; x_0))^{-1}dt \text{ exists}, \\ 0, & \text{otherwise}, \end{cases}$$

for an appropriately selected function $g : \mathbb{R}^n \to \mathbb{R}$ satisfying the bounds $\gamma_1(|x|) \le g(x) \le \gamma_2(|x|)$ for all $x \in \mathbb{R}^n$ for $\mathcal{K}_\infty$-functions $\gamma_1, \gamma_2 \in \mathcal{K}_\infty$. Using J, the optimal value function defined as

$$C(x_0) = \sup_{\phi \in \mathcal{S}(x_0)} J(x_0, \phi) \quad \Longleftrightarrow \quad \frac{1}{C(x_0)} = \inf_{\phi \in \mathcal{S}(x_0)} \frac{1}{J(x_0, \phi)}$$

for $x_0 \ne 0$ is shown to be continuous. Under the assumption that $C(x_0) = J(x_0, \psi)$ for $\psi \in \mathcal{S}(x_0)$, the dynamic programming principle guarantees

$$\frac{1}{C(x_0)} = \int_0^T \frac{1}{g(\psi(t; x_0))}dt + \frac{1}{C(\psi(T; x_0))}$$

for $T \in \mathbb{R}_{\geq 0}$. Rearranging terms, dividing by $T > 0$ and considering the limit $T \to 0$ leads to

$$0 = \frac{1}{g(\psi(0; x_0))} - \frac{1}{C(\psi(0; x_0))^2} D^+ C(x_0; w)$$

for a $w \in F(x)$, from which the increase condition (4.8) is derived.

Theorem 4.5 suggests that Theorem 4.11 can be strengthened to the existence of Lipschitz continuous control Chetaev functions. To prove that the existence of a continuous control Chetaev function implies weak $\mathcal{K}_\infty\mathcal{K}_\infty$-instability we indeed make use a control Chetaev function which is Lipschitz continuous excluding a neighborhood around the origin. To this end, we first show that the existence of a continuous control Chetaev function implies the existence of a continuous control Chetaev function which is Lipschitz continuous for all $x \in \mathbb{R}^n$ satisfying $\delta \leq |x| \leq \Delta$ for arbitrary parameters $\Delta > \delta > 0$ (see Lemma 6.14). We do not attempt to construct a control Chetaev function which is Lipschitz continuous for all $x \in \mathbb{R}^n$ here. Instead, we will show that allowing for nonsmooth control Chetaev functions is indeed necessary, i.e., that there are differential inclusions (2.1) satisfying Assumptions 2.1 and 2.2, which are weakly completely unstable according to Definition 4.6 and which admit a control Chetaev function in the Dini sense but no smooth control Chetaev function.

In the context of stability and control Lyapunov functions, the nonlinear control systems known as Artstein's circles [5] or the Brockett integrator [16] are the standard examples for systems that are weakly $\mathcal{K}\mathcal{L}$-stable but for which smooth control Lyapunov functions do not exist. In the context of weak complete instability we can even use a linear system as an example.

Example 4.12 Consider the differential inclusion

$$\dot{x} \in F(x) = \overline{\mathrm{conv}}\{f(x, u) | u \in \mathcal{U}(x)\} \tag{4.9}$$

where $f(x, u)$ and $\mathcal{U}$ are defined as

$$f(x, u) = \begin{bmatrix} 1 & 0 \\ 0 & -1 \end{bmatrix} x + \begin{bmatrix} 1 \\ 0 \end{bmatrix} u \quad \text{and} \quad \mathcal{U}(x) = [-2|x|, 2|x|]. \tag{4.10}$$

The same argument regarding smooth Lyapunov and Chetaev functions for time-reversed systems presented in Corollary 3.18 can be applied for continuous stabilizers and destabilizers and control Lyapunov and control Chetaev functions (see Corollary 4.19 below).

In the context of (4.9), (4.10), assume there exist a smooth control Chetaev function C and a function $\rho \in \mathcal{P}$ such that

$$\sup_{u \in \mathcal{U}(x)} \langle \nabla C(x), f(x, u) \rangle \geq \rho(|x|) \quad \Longleftrightarrow \quad \min_{u \in \mathcal{U}(x)} \langle \nabla C(x), -f(x, u) \rangle \leq -\rho(|x|).$$

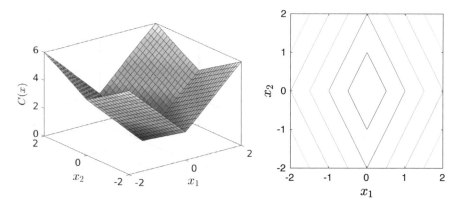

Fig. 4.1 Visualization of the nonsmooth candidate control Chetaev functions

Then $V = C$ is a control Lyapunov function for the dynamics $\dot{x} = -f(x, u)$. However, the second component x_2 of $-f$, is clearly not stabilizable to the origin, implying that a smooth control Lyapunov function cannot exist. Consequently, a smooth control Chetaev function for (4.9), (4.10) cannot exist. Note that the argument here crucially relies on the smoothness of the function C, which allows the if and only if statement above. By contrast, as discussed in Section 2.3 this time reversal fails to work when considering nonsmooth functions and Dini derivatives.

Nevertheless, intuitively it is plausible that the origin is weakly completely unstable. For $x_1(0) \neq 0$ and $u(t) = 0$ for $t \geq 0$, $\dot{x}_1 = x_1$ is satisfied and thus $|\phi(t; x_0)| \to \infty$ for $t \to \infty$ independent of the selection of $x_2(0)$. If $x_1(0) = 0$, the degree of freedom in u can be used to push the solution from the x_2-axis. Thus, it only needs to be shown that (4.4) is satisfied to conclude weak complete instability of the origin.

Consider the candidate control Chetaev function

$$C(x) = 2|x_1| + |x_2| \qquad (4.11)$$

which satisfies

$$\alpha_1(|x|) = |x| \leq C(x) = 2|x_1| + |x_2| \leq 2\sqrt{2}|x| = \alpha_2(|x|)$$

and which is shown in Fig. 4.1.

Moreover, $C(x)$ is Lipschitz continuous and satisfies

$$\nabla C(x) = \begin{bmatrix} 2\,\text{sign}(x_1) \\ \text{sign}(x_2) \end{bmatrix}, \qquad \forall\, x \in \mathbb{R}^2 \setminus (\{0\} \times \mathbb{R} \cup \mathbb{R} \times \{0\})\,.$$

On the x_1-x_2-axis the Dini derivative satisfies

$$D^+C(x; w) = 2\,\mathrm{sign}(x_1)w_1 + |w_2| = |w_2|, \qquad \forall\, x \in \{0\} \times \mathbb{R}\backslash\{0\},$$
$$D^+C(x; w) = 2|w_1| + \mathrm{sign}(x_2)w_2 = 2|w_1|, \qquad \forall\, x \in \mathbb{R}\backslash\{0\} \times \{0\},$$

and $D^+V(0; w) = 2|w_1| + |w_2|$ at the origin.

To show that C satisfies (4.8) and thus is a control Chetaev function we consider the feedback law

$$u(x) = s(x_1)|x_2| \in \mathcal{U}(x) \qquad \forall x \in \mathbb{R}^2 \tag{4.12}$$

where $s : \mathbb{R} \to \mathbb{R}$ is a particular definition of the sign-function

$$s(r) = \begin{cases} 1, \; r \geq 0, \\ -1, \; r < 0. \end{cases}$$

Note that in contrast to the standard definition of the sign-function, $s(0) \neq 0$. In what follows, we could have alternatively defined $s(0) = -1$ to obtain the same qualitative results. However, a discontinuous feedback law is necessary to ensure that $|\phi(t; x_0)| \to \infty$ for all $x_0 \in \mathbb{R}^2\backslash\{0\}$. In particular, since $|\phi_1(t; x_0)| \to 0$ for all $x_0 \in \mathbb{R}^2$, a discontinuous decision with respect to the feedback law is necessary which ensures that $\phi_1(t; x_0) \to \infty$ or $\phi_1(t; x_0) \to -\infty$ depending on the initial state. Thus, at best a Lipschitz continuous (but not a continuously differentiable) control Chetaev function can exist.

With the definitions so far and with $\alpha_3(r) = r$ it holds that

$$\begin{aligned} D^+C(x; f(x, u(x))) &= \langle \nabla C(x), f(x, u(x)) \rangle \\ &= 2\,\mathrm{sign}(x_1)(x_1 + s(x_1)|x_2|) + \mathrm{sign}(x_2)(-x_2) \\ &= 2|x_1| + 2|x_2| - |x_2| \geq \alpha_3(|x|) \end{aligned}$$

for all $x \in \mathbb{R}^2\backslash (\{0\} \times \mathbb{R} \cup \mathbb{R} \times \{0\})$, as well as

$$D^+C(x; f(x, u(x))) = |-x_2| = |x_2| \geq \alpha_3(|x|) \qquad \forall\, x \in \{0\} \times \mathbb{R}\backslash\{0\},$$
$$D^+C(x; f(x, u(x))) = 2|(x_1 + s(x_1)|x_2|)| = 2|x_1| \geq \alpha_3(|x|) \qquad \forall\, x \in \mathbb{R}\backslash\{0\} \times \{0\},$$

and

$$D^+C(0; f(x, u(x))) = 2|(x_1 + s(x_1)|x_2|)| + |-x_2| = 0 \geq \alpha_3(0).$$

Thus C is a control Chetaev function according to Definition 4.10.

While we know from Theorem 4.11 that the existence of a control Chetaev function implies weak $\mathcal{K}_\infty\mathcal{K}_\infty$-instability according to Definition 4.8, due to the simplicity of the dynamics (4.12), an explicit bound (4.6) in terms of a $\mathcal{K}_\infty\mathcal{K}_\infty$-function can be derived here. It holds that $\dot{x}_2 = -x_2$ which implies

$$\phi_2(t; x_2(0)) = e^{-t} x_2(0)$$

and thus the solution of the first component can be derived as

$$\phi_1(t; x_1(0)) = \left(x_1(0) + \tfrac{s(x_1(0))}{2} |x_2(0)| \right) e^t - \tfrac{s(x_1(0))}{2} |x_2(0)| e^{-t}$$
$$= x_1(0) e^t + \tfrac{s(x_1(0))}{2} |x_2(0)| \left(e^t - e^{-t} \right).$$

Moreover, the solutions satisfy the estimate

$$|\phi(t; x_0)|^2 = \left| \left(x_1(0) + \tfrac{s(x_1(0))}{2} |x_2(0)| \right) e^t - \tfrac{s(x_1(0))}{2} |x_2(0)| e^{-t} \right|^2 + |x_2(0) e^{-t}|^2$$

$$\geq \left(\left| \left(x_1(0) + \tfrac{s(x_1(0))}{2} |x_2(0)| \right) e^t \right| - \left| \tfrac{s(x_1(0))}{2} |x_2(0)| e^{-t} \right| \right)^2 + |x_2(0) e^{-t}|^2$$

$$= \left(\left| \left(x_1(0) + \tfrac{s(x_1(0))}{2} |x_2(0)| \right) e^t \right| - \tfrac{1}{2} |x_2(0)| e^{-t} \right)^2 + |x_2(0) e^{-t}|^2$$

$$= \left(x_1(0) + \tfrac{s(x_1(0))}{2} |x_2(0)| \right)^2 e^{2t} + \tfrac{1}{4} x_2(0)^2 e^{-2t}$$
$$\quad - |x_1(0)||x_2(0)| - \tfrac{1}{2} x_2(0)^2 + x_2(0)^2 e^{-2t}$$

$$= x_1(0)^2 e^{2t} + |x_1(0)||x_2(0)| e^{2t} + \tfrac{1}{4} x_2(0)^2 e^{2t} + \tfrac{5}{4} x_2(0)^2 e^{-2t}$$
$$\quad - |x_1(0)||x_2(0)| - \tfrac{1}{2} x_2(0)^2$$

$$\geq x_1(0)^2 e^{2t} + x_2(0)^2 \left(e^{2t} + \tfrac{1}{4} e^{-2t} - \tfrac{1}{2} + e^{-2t} \right).$$

It holds that $1 + 2e^{-\frac{1}{2}t} \leq 3$ and $e^{\frac{3}{2}t} + 5e^{-\frac{5}{2}t} \geq 3$. The second inequality can be shown by calculating the minimum over $t \in [0, \infty)$. Thus $e^{\frac{3}{2}t} + 5e^{-\frac{5}{2}t} \geq 1 + 2e^{-\frac{1}{2}t}$ for all $t \in [0, \infty)$ or equivalently

$$\tfrac{1}{4} e^{2t} + \tfrac{5}{4} e^{-2t} - \tfrac{1}{2} \geq \tfrac{1}{4} e^{\frac{1}{2}t} \qquad \forall \, t \geq 0.$$

If we use this estimate in the expression derived before, it holds that

$$|\phi(t; x_0)|^2 \geq x_1(0)^2 e^{2t} + \tfrac{1}{4} x_2(0)^2 e^{\frac{1}{2}t} \geq \tfrac{1}{4} \left(x_1(0)^2 + x_2(0)^2 \right) e^{\frac{1}{2}t}.$$

Thus (4.9) is weakly completely unstable according to Definition 4.8 with $\mathcal{K}_\infty \mathcal{K}_\infty$-function $\kappa(r, t) = \tfrac{1}{2} r e^{\frac{1}{\sqrt{2}} t}$.

In Fig. 4.2 four solutions of (4.9) using the feedback law (4.12) are visualized together with the level sets of the nonsmooth control Chetaev function (4.11).

Figure 4.3 shows $|\phi(t; x_0)|$ and $\kappa(|x_0|, t)$ for the four initial conditions used in Fig. 4.2. While the norm of the solutions is not strictly monotonically increasing, the $\mathcal{K}_\infty \mathcal{K}_\infty$-function κ provides a strictly monotonically increasing lower bound.

Following the arguments of this counterexample we have shown the result stated next.

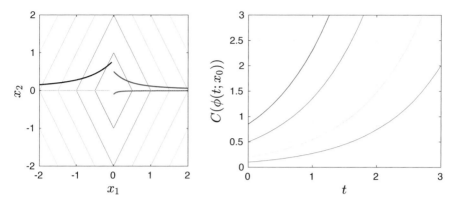

Fig. 4.2 On the left, solutions of (4.9) in the x_1-x_2-plane together with the level sets of the function (4.11). On the right, the monotonicity of $C(\phi(t; x_0))$ along solutions indicates that C is indeed a control Chetaev functions

Fig. 4.3 Visualization of $\phi(t; x_0)$ (solid lines) and $\kappa(|x_0|, t)$ (solid lines with crosses) for four initial values. Graphs for identical initial values have identical colors. As expected, $|\phi(t; x_0)| \geq \kappa(|x_0|, t)$ is satisfied for all $t \geq 0$

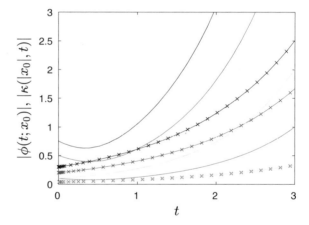

Corollary 4.13 *There are differential inclusions satisfying Assumptions 2.1 and 2.2 which are weakly $\mathcal{K}_\infty \mathcal{K}_\infty$-unstable and which do not admit smooth control Chetaev functions.*

4.3 Relations Between Control Chetaev Functions, Control Lyapunov Functions, and Scaling

While the results derived in Sect. 3.3 based on the scaled differential inclusions $\dot{x} \in \eta(|x|)F(x)$ and Theorem 2.3 remain valid in the context of weak $\mathcal{K}\mathcal{L}$-stability and weak $\mathcal{K}_\infty \mathcal{K}_\infty$-instability, the connections between $\dot{x} \in F(x)$ and $\dot{x} \in -F(x)$ cannot be copied in a similar way.

In particular, a result such as Corollary 3.18 cannot be derived in the context of nonsmooth control Lyapunov/Chetaev functions. To see this, assume that V is a control Lyapunov function for the differential inclusion (2.1), i.e., the condition

$$-\rho(|x|) \geq \min_{w \in F(x)} D_+ V(x; w)$$

holds for an appropriate function $\rho \in \mathcal{P}$ for all $x \in \mathbb{R}^n$. Using the definition of the lower right Dini derivative, this condition can be equivalently written as

$$
\begin{aligned}
\rho(|x|) \leq \max_{w \in F(x)} -D_+ V(x; w) &= \max_{w \in F(x)} \left(-\liminf_{v \to w;\ t \searrow 0} \tfrac{1}{t}(V(x + tv) - V(x)) \right) \\
&= \max_{w \in F(x)} \limsup_{v \to w;\ t \searrow 0} -\tfrac{1}{t}(V(x + tv) - V(x)) \\
&= \max_{w \in F(x)} \limsup_{v \to w;\ t \nearrow 0} \tfrac{1}{t}(V(x - tv) - V(x)) \\
&= \max_{w \in -F(x)} \limsup_{v \to w;\ t \nearrow 0} \tfrac{1}{t}(V(x + tw) - V(x)) = \max_{w \in -F(x)} D^- V(x; w).
\end{aligned}
$$

$$(4.13)$$

The calculations above show that the right Dini derivative becomes a left Dini derivative, which cannot be used to determine an increase of the Chetaev function in forward time for the time reversal system (2.2), cf. the discussion at the end of Sect. 2.3. The same arguments hold if we start with a nonsmooth control Chetaev function instead of a nonsmooth control Lyapunov function.

The fact that the existence of a nonsmooth control Lyapunov function for $\dot{x} \in F(x)$ indeed does not imply that there exists a nonsmooth control Chetaev function for $\dot{x} \in -F(x)$ can be observed on the example of Artstein's circles [5].

Example 4.14 (*Artstein's circles*) The dynamical system $\dot{x} = f(x, u)$ described by

$$\dot{x}_1(t) = \left(-x_1(t)^2 + x_2(t)^2\right) u(t), \qquad \dot{x}_2(t) = (-2x_1(t)x_2(t)) u(t) \qquad (4.14)$$

is known as Artstein's circles in the literature. For $u \in [-1, 1] = \mathcal{U}$ and $F(x) = \overline{\text{conv}}\{f(x, u) | u \in \mathcal{U}\}$ the dynamics can be described in the form of a differential inclusion (2.1). The function

$$V(x) = \sqrt{4x_1^2 + 3x_2^2} - |x_1| \qquad (4.15)$$

is a control Lyapunov function in the Dini sense according to Definition 4.4 (see [5]), which implies weak $\mathcal{KL}$-stability according to Theorem 4.5. The control Lyapunov function is shown in Fig. 4.4.

The time reversal system is not weakly completely unstable since all solutions of Artstein's circles with initial value $x \in \mathbb{R}^2 \backslash (\mathbb{R} \times \{0\})$ are bounded for all $t \in \mathbb{R}_{\geq 0}$.

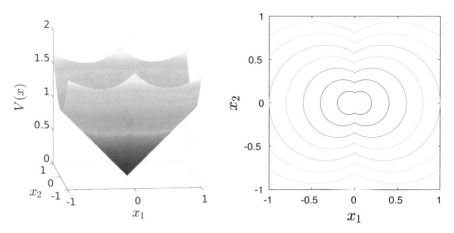

Fig. 4.4 Nonsmooth control Lyapunov function (4.15) for the dynamics (4.14). The function V is nonsmooth on the x_2-axis

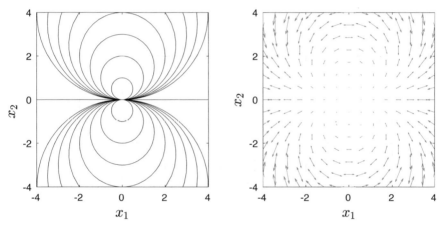

Fig. 4.5 On the left, solutions of the dynamics (4.14) for different initial conditions and $u \in \{-1, 1\}$. On the right, phase portrait of (4.14) for $u = 1$ (blue) and $u = -1$ (red)

More explicitly, all solutions of the dynamical system (4.14) are described through circles, where the radius of the circle is defined by the initial value. The input u can only change the direction (left or right) and the velocity of the solution. Fig. 4.5 shows solutions of (4.14) for different initial conditions (left) and the phase portrait (right) with respect to constant input signals $u = -1$ and $u = 1$, respectively.

For any potential control Chetaev function C there needs to exists at least one point $\tilde{x} \in \mathbb{R}^n$ on a circle corresponding to any initial value $x_0 \in \mathbb{R}^2 \backslash (\mathbb{R} \times \{0\})$ where no increasing direction $D^+C(\tilde{x}; w) > 0$ exists. This is true for initial values arbitrarily close to the origin and thus, also local arguments cannot be applied.

The Dini derivatives of (4.15) for $x \in \{0\} \times (\mathbb{R} \backslash \{0\})$ in direction $w \in \mathbb{R}^2$ satisfy

$$D_+V(x, w) = D^+V(x, w) = \frac{1}{\sqrt{3x_2^2}}\begin{bmatrix} 0 & 3x_2 \end{bmatrix}\begin{bmatrix} w_1 \\ w_2 \end{bmatrix} - |w_1|$$

$$D_-V(x, w) = D^-V(x, w) = \frac{1}{\sqrt{3x_2^2}}\begin{bmatrix} 0 & 3x_2 \end{bmatrix}\begin{bmatrix} w_1 \\ w_2 \end{bmatrix} + |w_1|$$

In $x_1 = 0$, in direction $w = f(x, u)$, $u \in [-1, 1]$ it holds that

$$D_+V(x, f(x, u)) = D^+V(x, f(x, u)) = -|x_2^2 u| = -x_2^2 |u| \leq 0 \qquad (4.16)$$
$$D_-V(x, f(x, u)) = D^-V(x, f(x, u)) = |x_2^2 u| = x_2^2 |u| \geq 0 \qquad (4.17)$$

(since $\dot{x}_2 = 0$ for $x_1 = 0$). Thus, (4.16) confirms the illustration in Fig. 4.5 that on the x_2-axis $V(\phi(t; x_0))$ can only decrease. Moreover, as outlined in (4.13) the left Dini derivative (4.17) does not determine an increase of the Chetaev function in forward time.

We summarize this observation in the following corollary.

Corollary 4.15 *Let F satisfy Assumptions 2.1 and 2.2. Weak $\mathcal{KL}$-stability of the origin for $\dot{x} \in F(x)$ is not equivalent to weak $\mathcal{K}_\infty\mathcal{K}_\infty$-instability of the origin for $\dot{x} \in -F(x)$.*

The result shows that even though there are similarities between stability in forward time and instability in backward time, instability results cannot simply be obtained by mirroring known results from stability theory.

While Artstein's circles show that there are systems which are weakly $\mathcal{KL}$-stable in forward time but not necessarily weakly $\mathcal{K}_\infty\mathcal{K}_\infty$-unstable in backward time, the converse in terms of stability and instability can also be illustrated by means of an example. In particular, there are systems which are weakly $\mathcal{K}_\infty\mathcal{K}_\infty$-unstable in forward time which are not weakly $\mathcal{KL}$-stable in backward time.

Example 4.16 Recall the dynamics (4.9) defined through (4.10) and introduced in Example 4.12. In Example 4.12 we have seen that the system is weakly $\mathcal{K}_\infty\mathcal{K}_\infty$-unstable but does not admit a smooth control Chetaev function. Moreover, the system is weakly $\mathcal{KL}$-stable with smooth control Lyapunov function $V(x) = |x|^2$ and smooth stabilizing feedback law $u(x) = -2x_1$.

In backward time the differential inclusion $\dot{x} \in -F(x)$ where F is defined through

$$-f(x, u) = \begin{bmatrix} -1 & 0 \\ 0 & 1 \end{bmatrix} x + \begin{bmatrix} -1 \\ 0 \end{bmatrix} u \quad \text{and} \quad \mathcal{U}(x) = [-2|x|, 2|x|]$$

is weakly $\mathcal{K}_\infty\mathcal{K}_\infty$-unstable with respect to the origin and with control Chetaev function $C(x) = |x|^2$ (and smooth feedback law $u(x) = -2x_1$). However, $\dot{x} \in -F(x)$ is not weakly $\mathcal{KL}$-stable since $|\psi_2(t; x_0)| \to \infty$ for all $x_0 \in \mathbb{R}^2\backslash\{0\}$ for all $u \in \mathcal{U}(x)$.

From the description of the solutions of Artstein's circles in Example 4.14 one can deduce that the system (4.14) is stabilizable but not controllable, i.e., it is possible to reach the origin from any initial position (in finite time), but it is not possible to reach an arbitrary point in $\mathbb{R}^2$ from an arbitrary initial position.

Similarly, the dynamics (4.9) are stabilizable but not controllable, and thus the lack of controllability demonstrates one scenario where weak $\mathcal{K}_\infty \mathcal{K}_\infty$-instability cannot be deduced from weak $\mathcal{KL}$-stability by looking at the dynamics in backward time.

Note however that also for controllable systems (2.1), it is not sufficient to construct control Lyapunov functions for the time-reversed dynamics (2.2) to obtain a control Chetaev function for (2.1). We illustrate this property on the dynamics of the Brockett integrator, which is known to be controllable but which does not admit a smooth control Lyapunov function.

Example 4.17 Consider the dynamics of the Brockett integrator [16],

$$F(x) = \overline{\mathrm{conv}}\{f(x, u)|u \in \mathcal{U}\} \tag{4.18}$$

defined through

$$f(x, u) = \begin{bmatrix} u_1 \\ u_2 \\ x_1 u_2 - x_2 u_1 \end{bmatrix} \quad \text{and} \quad \mathcal{U} = [-1, 1]^2 . \tag{4.19}$$

It follows from the function $f(x, u)$ and the symmetry of the set $\mathcal{U}$ with respect to the origin that the dynamics in forward time are equivalent to the dynamics in backward time. In [20] it is shown that

$$V(x) = x_1^2 + x_2^2 + 2x_3^2 - 2|x_3|\sqrt{x_1^2 + x_2^2} \tag{4.20}$$

is a Lipschitz continuous control Lyapunov function for (4.19) according to Definition 4.4 (see in particular [20, Page 23]).

However, (4.20) is not a control Chetaev function for the (time reversal) system. To see this, we compute the right Dini derivative of $V(x)$ for $x_1 = x_2 = 0$ and $x_3 \neq 0$. (The lower right and the upper right Dini derivative of V coincide.) For $x_1 = x_2 = 0$ and $x_3 \neq 0$ it holds that

$$D_+V(x, w) = 2x_3 w_3 - 2 \lim_{t \searrow 0} \inf \frac{1}{t} \left(|x_3 + t w_3| \sqrt{w_1^2 t^2 + w_2^2 t^2} \right)$$

$$= 2x_3 w_3 - 2|x_3|\sqrt{w_1^2 + w_2^2}.$$

With $w = f(x, u)$ this implies

$$D_+V(x, f(x, u)) = 2x_3(x_1u_2 - x_2u_1) - 2|x_3|\sqrt{u_1^2 + u_2^2}$$

$$= -2|x_3|\sqrt{u_1^2 + u_2^2} \leq 0$$

for all $x \in \{0\} \times \{0\} \times \mathbb{R}\backslash\{0\}$ and for all $u \in \mathcal{U}$, i.e., an increasing direction with respect to V does not exist and V is not a control Chetaev function as claimed.

Nevertheless, a nonsmooth control Chetaev function according to Definition 4.10 for (4.20) can be defined.

Claim 4.18 *The function* $C : \mathbb{R}^3 \to \mathbb{R}_{\geq 0}$, $C(x) = |x_1| + |x_2| + |x_3|$ *defines a Lipschitz continuous control Chetaev function according to Definition 4.10 for* (4.18), (4.19).

Proof of Claim 4.18 The proof is not difficult, but tedious. Lipschitz continuity of C follows from the Lipschitz continuity of the absolute value function.

The lower and the upper bound (4.7) are satisfied for $\alpha_1(|x|) = |x|$ and $\alpha_2(|x|) = \sqrt{3}|x|$ which follows from the norm equivalence of the 1-norm and the 2-norm.

For $x_1 \neq 0$, $x_2 \neq 0$ and $x_3 \neq 0$, the function C is continuously differentiable and satisfies

$$\langle \nabla C(x), f(x, u) \rangle = \left\langle \begin{bmatrix} \text{sign}(x_1) \\ \text{sign}(x_2) \\ \text{sign}(x_3) \end{bmatrix}, \begin{bmatrix} u_1 \\ u_2 \\ x_1u_2 - x_2u_1 \end{bmatrix} \right\rangle$$

$$= (\text{sign}(x_1) - \text{sign}(x_3)x_2)u_1 + (\text{sign}(x_2) + \text{sign}(x_3)x_1) + u_2.$$

By checking the cases

- $\text{sign}(x_1) = 1$, $\text{sign}(x_3) = 1$, $x_2 = 1$,
- $\text{sign}(x_1) = -1$, $\text{sign}(x_3) = 1$, $x_2 = -1$,
- $\text{sign}(x_1) = -1$, $\text{sign}(x_3) = -1$, $x_2 = 1$, and
- $\text{sign}(x_1) = 1$, $\text{sign}(x_3) = -1$, $x_2 = -1$,

it can be verified that

$$(\text{sign}(x_2) + \text{sign}(x_3)x_1) \neq 0 \quad \text{whenever} \quad (\text{sign}(x_1) - \text{sign}(x_3)x_2) = 0.$$

Hence, there exists $u \in [-1, 1]^2$ such that $D^+C(x; f(x, u)) > 0$ for all $x_1 \neq 0$, $x_2 \neq 0$ and $x_3 \neq 0$.

For the remaining cases for x we rely on the Dini derivative.

- Let $x_1 = 0$, $x_2 \neq 0$, $x_3 \neq 0$ and $u_1 = 0$, $u_2 = \text{sign}(x_2)$. Then, it holds that

$$D^+C(x; f(x, u)) = |u_1| + \text{sign}(x_2)u_2 + \text{sign}(x_3)(-x_2u_1) = 1 > 0.$$

- Let $x_1 \neq 0$, $x_2 = 0$, $x_3 \neq 0$ and $u_1 = \text{sign}(x_1)$, $u_2 = 0$. Then, it holds that

$$D^+C(x; f(x, u)) = \text{sign}(x_1)u_1 + |u_2| + \text{sign}(x_3)(x_1u_2) = 1 > 0.$$

- Let $x_1 \neq 0$, $x_2 \neq 0$, $x_3 = 0$ and $u_1 = -\text{sign}(x_2)$, $u_2 = \text{sign}(x_1)$. Then, it holds that

$$D^+C(x; f(x, u)) = \text{sign}(x_1)u_1 + \text{sign}(x_2)u_2 + |x_1u_2 - x_2u_1|$$
$$= -\text{sign}(x_1)\text{sign}(x_2) + \text{sign}(x_2)\text{sign}(x_1) + |x_1\text{sign}(x_1) + x_2\text{sign}(x_2)|$$
$$= |x_1| + |x_2| > 0.$$

- Let $x_1 = 0$, $x_2 = 0$, $x_3 \neq 0$ and $u_1 = 1$, $u_2 = 0$. Then, it holds that

$$D^+C(x; f(x, u)) = |u_1| + |u_2| = 1 > 0.$$

- Let $x_1 = 0$, $x_2 \neq 0$, $x_3 = 0$ and $u_1 = 0$, $u_2 = \text{sign}(x_2)$. Then, it holds that

$$D^+C(x; f(x, u)) = |u_1| + \text{sign}(x_2)u_2 + |-x_2u_1| = 1 > 0.$$

- Let $x_1 \neq 0$, $x_2 = 0$, $x_3 = 0$ and $u_1 = \text{sign}(x_1)$, $u_2 = 0$. Then, it it holds that

$$D^+C(x; f(x, u)) = \text{sign}(x_1)u_1 + |u_2| + |x_1u_2| = 1 > 0.$$

- Let $x_1 = 0$, $x_2 = 0$, $x_3 = 0$ and $u_1 = 0$, $u_2 = 0$. Then, it holds that

$$D^+C(x; f(x, u)) = |u_1| + |u_2| = 0.$$

Thus, we can conclude that C is a control Chetaev function. □

While the consideration of nonsmooth control Lyapunov/Chetaev functions is necessary for some systems discussed in this section, if a smooth a control Lyapunov/Chetaev function exists, then a result similar to Corollary 3.18 immediately follows.

Corollary 4.19 *Consider the differential inclusion (2.1) satisfying Assumptions 2.1 and 2.2 together with its time reversed counterpart (2.2). The differential inclusion (2.1) admits a smooth control Lyapunov (Chetaev) function V (C) if and only if its time reversed counterpart (2.2) admits a smooth control Chetaev (Lyapunov) function $C = V$ ($V = C$).*

Proof The proof follows immediately from the proof of Corollary 3.18 by replacing min with max. □

As an example, controllable linear systems admit smooth control Lyapunov functions and smooth control Chetaev functions. As a final remark and a final result in this section we point out that the statements in Remark 3.17 and in Lemma 3.16 remain valid in the context of weak $\mathcal{KL}$-stability and weak $\mathcal{K}_\infty\mathcal{K}_\infty$-instability.

Remark 4.20 The scaling discussed in Remark 3.17 remains valid in the context of weak $\mathcal{KL}$-stability and weak $\mathcal{K}_\infty\mathcal{K}_\infty$-instability. In particular, (4.8) and (4.3) can be replaced by

$$\max_{w \in F(x)} D^+ C(x; w) \geq C(x), \tag{4.21}$$

$$\min_{w \in F(x)} D_+ V(x; w) \leq -V(x), \tag{4.22}$$

respectively.

Lemma 4.21 *Consider the differential inclusion (2.1) satisfying Assumptions 2.1 and 2.2 together with its scaled version (2.3) for a Lipschitz continuous positive scaling $\eta : \mathbb{R}_{\geq 0} \to \mathbb{R}_{>0}$.*

- *Assume that $V : \mathbb{R}^n \to \mathbb{R}_{\geq 0}$ is a continuous control Lyapunov function for (2.1) according to Definition 3.4. Then V is a continuous control Lyapunov function of the scaled differential inclusion (2.3).*
- *Assume that $C : \mathbb{R}^n \to \mathbb{R}_{\geq 0}$ is a continuous control Chetaev function for (2.1) according to Definition 3.14. Then C is a continuous control Chetaev function of the scaled differential inclusion (2.3).*

Proof The proof follows the same arguments as the proof of Lemma 3.16 by replacing the directional derivative with the Dini derivative. In particular let V denote a control Lyapunov function for (2.1). Then there exists $\rho \in \mathcal{P}$ such that the inequality

$$\min_{w \in F(x)} D_+ V(x; w) \leq -\rho(|x|) \tag{4.23}$$

is satisfied for all $x \in \mathbb{R}^n$. Using the definition of the Dini derivative, for all $x \in \mathbb{R}^n$ it holds that

$$\begin{aligned}
\min_{w \in \eta(|x|)F(x)} D_+ V(x; w) &= \min_{w \in \eta(|x|)F(x)} \liminf_{v \to w;\, t \searrow 0} \frac{1}{t} \left(V(x + tv) - V(x) \right) \\
&= \min_{w \in F(x)} \liminf_{v \to w\eta(|x|);\, t \searrow 0} \frac{1}{t} \left(V(x + tv) - V(x) \right) \\
&= \min_{w \in F(x)} \liminf_{v \to w;\, t \searrow 0} \frac{1}{t} \left(V(x + tv\eta(|x|)) - V(x) \right) \\
&= \min_{w \in F(x)} \liminf_{v \to w;\, t \searrow 0} \frac{\eta(|x|)}{t} \left(V(x + tv) - V(x) \right) \\
&= \eta(|x|) \min_{w \in F(x)} D_+ V(x; w) \leq -\eta(|x|)\rho(|x|).
\end{aligned}$$

The remaining arguments are the same as in the proof of Lemma 3.16.   □

4.4 Comparison to Control Barrier Function Results

Another approach using Lyapunov like arguments, which became quite popular during the last years but is out of the scope of this monograph is based on *barrier* and *control barrier functions* (see [1, 2], for example).

Control barrier functions are usually defined for nonlinear control affine systems

$$\dot{x} = f(x) + g(x)u, \tag{4.24}$$

with state $x \in \mathbb{R}^n$, input $u \in \mathcal{U} \subset \mathbb{R}^m$ and locally Lipschitz continuous functions $f : \mathbb{R}^n \to \mathbb{R}^n$, $g : \mathbb{R}^{n \times m} \to \mathbb{R}^n$. Instead of (in)stability or (de)stabilization, barrier functions or control barrier functions are used to characterize *safety* in terms of forward invariance.

A set $C \subset \mathbb{R}^n$ is called forward invariant if for every $x_0 \in C$,

$$\phi(t; x_0) \in C, \qquad \forall t \in \mathbb{R}_{\geq 0}.$$

Similar to the distinction in Chaps. 3 and 4 forward invariance can be understood in the strong or in the weak sense, i.e., required for all solutions $\phi \in S(x_0)$ or required only for at least one solution $\phi \in S(x_0)$.

For a Lipschitz continuous feedback law $u = k(x), k : \mathbb{R}_{\geq 0} \to \mathcal{U}$, the closed loop system $\dot{x} = f(x) + g(x)k(x)$ is called safe with respect to C if C is forward invariant. This corresponds to the interpretation that the states in C are safe while the states outside are unsafe.

Definition 4.22 (*Control barrier function*) Let $C \subset \mathbb{R}^n$ be the superlevel set

$$C = \{x \in \mathbb{R}^n \mid B(x) \geq 0\}. \tag{4.25}$$

of a continuously differentiable function $B : \mathbb{R}^n \to \mathbb{R}$. Then B is a control barrier function if there exists an extended class $\mathcal{K}_\infty$ function $\delta : \mathbb{R} \to \mathbb{R}$ such that

$$\sup_{u \in \mathcal{U}} \left(\langle \nabla B(x), f(x) \rangle + \langle \nabla B(x), g(x) \rangle u \right) \geq -\delta(B(x)) \tag{4.26}$$

A function $\delta : \mathbb{R} \to \mathbb{R}$ is said to be an extended $\mathcal{K}_\infty$ function if there exist $\alpha_1, \alpha_2 \in \mathcal{K}_\infty$ so that $\delta(r) = \alpha_1(r)$ and $\delta(-r) = -\alpha_2(r)$ for all $r \in \mathbb{R}_{\geq 0}$.

If $B(x)$ is a control barrier function according to Definition 4.22 and C is defined in (4.25), then C is safe and asymptotically stable with respect to (4.24) and a control law $u = k(x)$ satisfying inequality (4.26), [1, Theorem 2].

Remark 4.23 In terms of control Lyapunov functions this can be interpreted as asymptotic stabilizability of a forward invariant set C with control Lyapunov function characterization

$$V(x) = \begin{cases} -B(x), & \text{for } B(x) < 0 \\ 0, & \text{for } B(x) \geq 0 \end{cases}$$

satisfying

$$\alpha_1(|x|_C) \leq V(x) \leq \alpha_2(|x|_C) \qquad \forall x \in \mathbb{R}^n$$

$$\inf_{u \in \mathcal{U}} D_+ V(x; f(x, u)) \leq -\alpha_3(V(x)) \qquad \forall x \in \mathbb{R}^n$$

for $\alpha_1, \alpha_2 \in \mathcal{K}_\infty$, $\alpha_3(r) = \delta(-r)$ and $|x|_C = \inf_{y \in C} |x - y|$.

Instead of characterizing asymptotic stability or complete instability properties of equilibria, which is the main focus of this book, the idea behind inequality (4.26) is that in the interior of C, where $B(x)$ is large, (4.26) is not restrictive when selecting u, whereas for $x \in \{x \in \mathbb{R}^n \mid B(x) = 0\}$ inequality (4.26) is restrictive and guarantees that the set C is not left if u is selected appropriately.

In the design of the controller $u = k(x)$, the property of the set C characterized through (4.26) is usually used as a constraint, accompanying a constraint based on a control Lyapunov function. In particular, based on a smooth control Lyapunov function V and a smooth control barrier function B a control law $u(x)$ for affine systems (4.24) can be implicitly defined through

$$
\begin{aligned}
u = k(x) = \operatorname*{argmin}_{\substack{u \\ (u,\gamma) \in \mathcal{U} \times \mathbb{R}}} \quad & u^T H(x) u + \gamma^2 \\
\text{subject to} \quad & \langle \nabla V(x), f(x) + g(x)u \rangle \leq -\rho(|x|) + \gamma \\
& \langle \nabla B(x), f(x) + g(x)u \rangle \geq -\delta(B(x)),
\end{aligned}
\tag{4.27}
$$

where argmin_u selects the u-component of the argmin. Here, $\rho \in \mathcal{P}$ denotes the positive used in (4.3) in Definition 4.4 and $H : \mathbb{R}^n \to \mathbb{R}^{m \times m}$ is a continuous function with $H(x)$ symmetric positive definite for all $x \in \mathbb{R}^n$. The first constraint is used to guarantee a decrease with respect to level sets of the control Lyapunov function while the second constraint ensures that the set C is not left. Since the two constraints might be conflicting, the slack variable γ is included to ensure feasibility of the optimization problem. Note, however, that (4.27) guarantees that the corresponding closed loop solution satisfies $\phi(t; x_0) \in C$ for all $t \in \mathbb{R}_{\geq 0}$, convergence $|\phi(t; x_0)| \to 0$ for $t \to \infty$ is not necessarily satisfied through the controller design (4.27).

Chapter 5
Outlook and Further Topics

Abstract In this chapter we briefly discuss two extensions of the results described in Chaps. 3 and 4, which combine stability and instability properties in a single framework or combine stability and instability through hybrid switching strategies. The two approaches described here, are based on the papers [13, 14].

Keywords Lyapunov methods · Differential inclusions · Stability of nonlinear systems · Instability of nonlinear systems · Stabilization/destabilization of nonlinear systems · Stabilizability and destabilizability

5.1 Complete Control Lyapunov Functions

Control Lyapunov functions guarantee weak $\mathcal{KL}$-stability of an equilibrium while control Chetaev functions guarantee weak $\mathcal{K}_\infty \mathcal{K}_\infty$-instability, i.e., avoidance of a neighborhood around an equilibrium. In [13], a possible extension unifying these two properties is discussed in terms of weak $\mathcal{KL}$-stability with avoidance properties and using *complete control Lyapunov functions*.

Definition 5.1 (*Weak $\mathcal{KL}$-stability with avoidance properties*) Let $O \subset \mathbb{R}^n, 0 \notin O$, be open. The differential inclusion (2.1) is weakly $\mathcal{KL}$-stable with respect to the origin with avoidance property with respect to O, if there exists $\beta \in \mathcal{KL}$ such that, for each $x_0 \in \mathbb{R}^n \backslash O$, there exists $\phi(\cdot; x_0) \in \mathcal{S}(x_0)$ so that

$$|\phi(t; x_0)| \leq \beta(|x_0|, t) \quad \text{and} \quad \phi(t; x_0) \notin O \quad \forall\, t \geq 0.$$

Trivially, in the case that $O = \emptyset$, Definition 5.1 reduces to Definition 4.2. A corresponding Lyapunov-like function for Definition 5.1 is given in terms of *complete control Lyapunov functions*. In what follows, we consider the special case that $O = \bigcup_{i=1}^{N} O_i$ for open sets $O_1, \ldots, O_N$.

© The Author(s), under exclusive license to Springer Nature Switzerland AG 2021
P. Braun et al., *(In-)Stability of Differential Inclusions*,
SpringerBriefs in Mathematics,
https://doi.org/10.1007/978-3-030-76317-6_5

Definition 5.2 ((*Complete control Lyapunov function*) Suppose that (2.1) satisfies Assumptions 2.1 and 2.2. For $i \in \{1, \ldots, N\}$, $N \in \mathbb{N}$, let $O_i \subset \mathbb{R}^n$ define open sets and let $V_C : \mathbb{R}^n \to \mathbb{R}$ be a continuous function. Assume there exist $\alpha_1, \alpha_2 \in \mathcal{K}_\infty$ and $\rho \in \mathcal{P}$ such that the following properties are satisfied. For all $i = \{1, \ldots, N\}$, there exist $c_i \in \mathbb{R}_{>0}$ such that

$$V_C(x) = c_i \quad \forall x \in \partial O_i \quad \text{and} \quad c_i \leq \inf_{x \in O_i} V_C(x). \tag{5.1}$$

Moreover,

$$\alpha_1(|x|) \leq V_C(x) \leq \alpha_2(|x|), \qquad \forall x \in \mathbb{R}^n, \tag{5.2}$$

and

$$\min_{w \in F(x)} D_+ V_C(x; w) \leq -\rho(x), \qquad \forall x \in \mathbb{R}^n \setminus \left(\bigcup_{i=1}^N O_i \right). \tag{5.3}$$

Then V_C is called complete control Lyapunov function.

Note that (5.2) implies that for all $i \in \{1, \ldots, N\}$ the boundary of O_i needs to be bounded. In particular, if ∂O_i is unbounded then $\alpha_1(|x|) \leq V(x) = c_i$ cannot be satisfied for all $x \in \partial O_i$. However, for this it is not necessary that O_i itself is bounded, as the example $O_i = \mathbb{R}^n \setminus \overline{B}_1(0)$ shows.

In [13] it is shown that the existence of a complete control Lyapunov function is sufficient for stability according to Definition 5.1.

Theorem 5.3 ([13, Theorem 2]) *Consider the differential inclusion (2.1) satisfying Assumptions 2.1 and 2.2. Additionally, let O_i, $i \in \{1, \ldots, N\}$, $N \in \mathbb{N}$, be open sets and let $V_C : \mathbb{R}^n \to \mathbb{R}$ be a complete control Lyapunov function according to Definition 5.2. Then the differential inclusion (2.1) is weakly $\mathcal{KL}$-stable with respect to the origin and has the avoidance property with respect to $O = \bigcup_{i=1}^N O_i$.*

As argued in [13], an immediate application of complete control Lyapunov functions is obtained by considering level sets of control Lyapunov functions. If V is a control Lyapunov function for (2.1), then for any $c \in \mathbb{R}_{>0}$ the superlevel set

$$O = \{x \in \mathbb{R}^n \mid V(x) > c\}$$

can be used as the open set in Definition 5.1. In this case the differential inclusion (2.1) is weakly $\mathcal{KL}$-stable with respect to the origin with avoidance property with respect to O according to Definition 5.1 since the sublevel set $O = \{x \in \mathbb{R}^n \mid V(x) \leq c\}$ is forward invariant and V is trivially a complete control Lyapunov function according to Definition 5.2.

A more interesting setting occurs when the open set O is bounded and, for simplicity, we assume that $N = 1$ holds in Definition 5.2. In this case, a possible complete control Lyapunov function is necessarily nonsmooth, which follows from the following result.

Theorem 5.4 ([12, Theorem 1]) *Let $O \subset \mathbb{R}^n$ be nonempty, open and bounded, and let $\alpha_1, \alpha_2 \in \mathcal{K}_\infty$ and $c \in \mathbb{R}_{>0}$. Let $W : \mathbb{R}^n \to \mathbb{R}_{\geq 0}$ be continuously differentiable[1] function satisfying the properties*

$$\alpha_1(|x|) \leq W(x) \leq \alpha_2(|x|), \tag{5.4}$$
$$W(x) = c \quad \forall\, x \in \partial O, \qquad W(x) > c \quad \forall\, x \in O, \tag{5.5}$$

and such that the union $\{x \in \mathbb{R}^n \mid W(x) \leq c\} \cup O$ is compact and path-connected[2]. Then, there exists $\bar{x} \in \{z \in \mathbb{R}^n \setminus \{0\} \mid W(z) \leq c\}$ such that $\nabla W(\bar{x}) = 0$.

Theorem 5.4 implies that the decrease condition (5.3) cannot hold at $\bar{x}$. Thus, a smooth complete control Lyapunov function cannot exist in the case that O is bounded. In [13], the continuously differentiable candidate complete control Lyapunov function

$$V_C(x) = |x| + 20C(x), \tag{5.6}$$

with

$$C(x) = \begin{cases} \frac{1}{2}\left(1 + \cos(|x - \hat{x}|^2)\right), & \text{for } |x - \hat{x}|^2 \leq \pi \\ 0, & \text{for } |x - \hat{x}|^2 > \pi \end{cases}$$

and $\hat{x} = [2, 0]^T$ is discussed. The function is visualized in Fig. 5.1.

From Theorem 5.4, and visually from Fig. 5.1, it follows that there exists $\bar{x} \in \mathbb{R}^n \setminus \{0, \hat{x}\}$ with $\nabla V_C(\bar{x}) = 0$. Thus, independent of the differential inclusion (2.1), V_C cannot be a complete control Lyapunov function according to Definition 5.2.

The problem described here, i.e., stabilizing the origin while simultaneously avoiding a neighborhood around $\hat{x} = [2, 0]^T$, is closely related to Artstein's circles discussed in Example 4.14 where a discrete decision is necessary to decide if the origin is approached from the left or the right. While in Example 4.14 the discrete decision is necessary due to topological obstructions induced by the dynamics, here the discrete decision results from a topological obstruction induced by the desired avoidance of O. In the context of control barrier functions and the controller design (4.27), the problem discussed here describes points $x \in \mathbb{R}^n$ where the constraints in (4.27) cannot be satisfied simultaneously with $\gamma = 0$.

A possible nonsmooth candidate complete control Lyapunov function is constructed for the problem of avoiding a neighborhood of the point $\hat{x} = [\frac{3}{2}, 0]^T$. With

---

[1]Different to the result in [12], a constant is added to the function W to be consistent with the notation in this monograph and to guarantee that $W(0) = 0$. The constant does not change the gradient of the function W and does not change the result.

[2]A path connecting $x, y \in X = \{x \in \mathbb{R}^n \mid W(x) \leq c\} \cup O$ is a continuous map $\gamma : [0, 1] \to X$, such that $\gamma(0) = x$ and $\gamma(1) = y$. Then X is path-connected if any two points $x, y \in X$ can be connected by a path. See [71, Def. 7.3.1], for example.

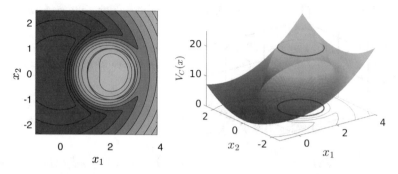

Fig. 5.1 Visualization of a continuously differentiable candidate complete control Lyapunov function (5.6). The red line indicates the boundary of the open set O. Intuitively it is clear that there needs to exist a point $\bar{x} \in \mathbb{R}^2 \backslash \{0, \hat{x}\}$ on the x_1-axis where the gradient satisfies $\nabla V_C(x) = 0$

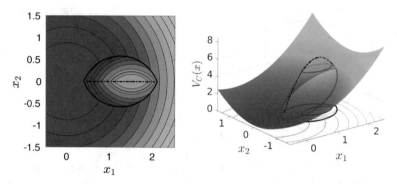

Fig. 5.2 Visualization of the function V_C defined in (5.7). The boundary of a potential open set O is visualized in red

$c_1 = [\frac{3}{2}, -\frac{2}{5}]^T, c_{-1} = [\frac{3}{2}, \frac{2}{5}]^T$ and parameters $\eta_1, \eta_2 \in \mathbb{R}_{>0}$ the functions

$$C_1(x) = -\eta_1 |x - c_1|^2 + \eta_2, \quad C_{-1}(x) = -\eta_1 |x - c_{-1}|^2 + \eta_2 \quad \text{and} \quad V(x) = |x|^2$$

are defined. For $\eta_1 = 5$ and $\eta_2 = 7$ these functions are combined through

$$V_C(x) = \max\{V(x), \min\{C_1(x), C_{-1}(x)\}\} \tag{5.7}$$

and shown in Fig. 5.2.

A numerical example relying on this construction guaranteeing $\mathcal{KL}$-stability while avoiding a neighborhood around a point $\hat{x}$ is discussed in [13].

Artificial potential fields and navigation functions (see [41, 42, 57], for example) are used to address a similar problem, i.e., asymptotically stabilizing the origin while avoiding bounded sets. These approaches rely on continuously differentiable functions, acknowledging that a decrease condition such as (5.3) can at best be guaranteed for a domain excluding a set of measure zero. This is comparable to the continuously differentiable function defined in (5.6) where the points $x \in \mathbb{R}^n$ satisfying $\nabla V_C(x) = 0$ describe a set of measure zero and would need to be excluded by assuming that the decrease condition is satisfied on the remaining set. For details we refer to the above references.

An alternative research stream addresses a related problem by characterizing stability properties of multistable systems through continuously differentiable Lyapunov functions combined with input-to-state stability properties [4, 31, 33]. Here, multistable systems potentially exhibit several equilibria, or more general forward invariant sets, and consequently the gradient of a corresponding continuously differentiable Lyapunov function not only vanishes at the origin but for every point on a forward invariant set. As an example consider the setting visualized in Fig. 5.1 with three forward invariant sets described through the minimum, the local maximum and the saddle point of the function V_C. However, global properties are still achieved in this research stream as in [33], for example, by leveraging input-to-state stability properties. More precisely, close to the saddle points, the nominal input derived from a continuously differentiable Lyapunov-like function is perturbed in a clever way, leading to a discontinuous feedback law guaranteeing that the saddle points are avoided and the origin is rendered globally attractive. The perturbation is defined based on input-to-state stability arguments ensuring that the perturbation does not affect the stability properties of the closed loop system defined through the nominal control law.

For a third relevant approach used to tackle the stability-with-avoidance problem based on control barrier functions, we refer to Sect. 4.4 for details.

5.2 Combined Stabilizing and Destabilizing Controller Design Using Hybrid Systems

Complete control Lyapunov functions are one possible extension to guarantee instability (or avoidance) properties of a set and stability properties of the origin. Alternatively one can derive a set of control laws based on control Lyapunov and control Chetaev functions discussed in Chap. 4 and orchestrate them in such a way that specific closed loop properties are satisfied. We illustrate this idea based on the results derived in [14], defining discontinuous feedback laws using a hybrid systems formalism. In particular, we use the hybrid systems to switch between individual control laws such that certain avoidance and stability properties are satisfied. For background on hybrid systems and for the notation used here, we refer to the monograph [32].

Instead of the differential inclusion (2.1) we consider the control system

$$\dot{x} = f(x, u), \qquad x_0 \in \mathbb{R}^n, \qquad u \in \mathcal{U},$$

directly, as motivated through the dynamics (2.8). Moreover, for $N \in \mathbb{N}$, we consider a set of feedback laws

$$u_q : \mathbb{R}^n \to \mathcal{U}, \qquad q \in \{1, \dots, N\}$$

and we define the extended state space $\Xi = \mathbb{R}^n \times \{1, \dots, N\}$ together with the extended state $\xi = [x^T, q]^T \in \Xi$. The additional discrete state is responsible for the controller selection which is coordinated through the *jump set* $\mathcal{D} = \cup_{i=1}^{N} \mathcal{D}_i \subset \Xi$, composed of closed sets $\mathcal{D}_i \subset \Xi$ for $i \in \{1, \dots, N\}$. The *flow set* C is defined as the closed complement of the jump set $C = \overline{\Xi \backslash \mathcal{D}}$.

With $\mathcal{D} \subset \Xi$, we define the set-valued map $G : \Xi \rightrightarrows \{1, \dots, N\}$,

$$G(x, q) = \big\{ i \in \{1, \dots, N\} | (x, q) \in \mathcal{D}_q, q \in \{1, \dots, N\} \big\}$$

returning an index $q \in \{1, \dots, N\}$ so that $(x, q) \in \mathcal{D}_q$ is satisfied. Note that the sets $\mathcal{D}_q, q \in \{1, \dots, N\}$, are allowed to overlap and thus G is indeed a set valued map.

With these definitions, the overall hybrid system is completed through the *flow dynamics* and the *jump dynamics*

$$\dot{\xi} = \begin{bmatrix} \dot{x} \\ \dot{q} \end{bmatrix} = \begin{bmatrix} f(x, u_q(x)) \\ 0 \end{bmatrix}, \qquad \xi \in C, \qquad (5.8)$$

$$\xi^+ = \begin{bmatrix} x^+ \\ q^+ \end{bmatrix} \in \begin{bmatrix} x \\ G(x, q) \end{bmatrix}, \qquad \xi \in \mathcal{D}. \qquad (5.9)$$

Here, the dynamics evolve according to a particular fixed controller selection in C until the set $\mathcal{D}$ is reached. Then, based on the function G, the controller is switched and the state proceeds according to the continuous dynamics (5.8) where the controller is unchanged. If the control laws u_q are derived using control Lyapunov and control Chetaev functions, avoidance and stability properties of the closed loop system (5.8)–(5.9) can be derived.

We illustrate these concepts on the example of the linear system

$$\dot{x} = \begin{bmatrix} 0 & 1 \\ -1 & 1 \end{bmatrix} x + \begin{bmatrix} 0 \\ 1 \end{bmatrix} u, \qquad u \in \mathbb{R}. \qquad (5.10)$$

discussed in [14]. As a setting, assume that we want to stabilize the origin, while simultaneously avoiding a neighborhood around the point $\hat{x} = [-1, 0]^T$.

Since (5.10) is a controllable linear system it is straightforward to find a control Lyapunov function and a feedback law u asymptotically stabilizing the origin. For example the quadratic function

$$V_1(x) = \frac{1}{2} x^T \begin{bmatrix} 3 & 1 \\ 1 & 1 \end{bmatrix} x$$

is a control Lyapunov function of (5.10) with respect to the origin and

$$u_1(x) = \begin{bmatrix} 0 & -3 \end{bmatrix} x$$

is a stabilizing feedback law. Here, u_1 and V_1 are obtained through pole placement and by solving a Lyapunov equation. Similarly, by shifting the origin,

$$V_2(x) = \frac{1}{2}(x - \hat{x})^T \begin{bmatrix} 3 & 1 \\ 1 & 1 \end{bmatrix} (x - \hat{x}) \quad \text{and} \quad u_2(x) = \begin{bmatrix} 0 & -3 \end{bmatrix} x - 1$$

define a control Lyapunov function with respect to $\hat{x}$ and u_2 renders $\hat{x}$ asymptotically stable. Finally,

$$C_3(x) = \frac{1}{2}(x - \hat{x})^T \begin{bmatrix} 3 & -1 \\ -1 & 1 \end{bmatrix} (x - \hat{x}) \quad \text{and} \quad u_3(x) = \begin{bmatrix} 0 & 1 \end{bmatrix} x - 1$$

define a control Chetaev function with respect to $\hat{x}$ and a corresponding destabilizing feedback law. The functions u_3 and C_3 are obtained through pole placement and by solving a Lyapunov equation corresponding to the shifted and time reversed system of (5.10).

A control law with avoidance guarantees is obtained by combining the three control laws u_1, u_2, and u_3 in the hybrid setting (5.8)–(5.9) and by defining the jump sets based on level sets of the functions V_1, V_2 and C_3. While V_2 and u_2 may not be necessary in the controller design, they are used in [14] to prove that the overall controller solves the combined avoidance and stabilization problem. In particular, while V_1 and u_1 guarantee asymptotic stability of the origin in the case that the obstacle is not present and C_3 and u_3 guarantee obstacle avoidance, V_2 and u_2 are included to ensure that a control law based on u_1, u_2, and u_3 does not introduce additional equilibria or periodic solutions. In Fig. 5.3 closed loop solutions using the hybrid setting (5.8)–(5.9) together with level sets of the functions V_1, V_2, and C_3 are shown.

Solutions starting inside the level set $C_3(x) \leq 0.1$ (red) start with the feedback law u_3 and the properties of the control Chetaev function guarantee that the set defined through $C_3(x) \leq 0.1$ is left. In the set defined through $V_2(x) \leq 0.6$ (cyan) and $C_3(x) \geq 0.1$, a convex combination of u_2 and u_3 is used to ensure that solutions are trapped inside the set $V_2(x) \leq 0.6$ and $C_3(x) \geq 0.1$ and eventually reach the set $V_1(x) \leq 0.7$. Outside the level set $V_2(x) \geq 0.7$, the stabilizing control law u_1 is used to guarantee asymptotic convergence to the origin. Once the solution satisfies $V_1(x) \leq 0.7$ (blue), forward invariance guarantees that solutions avoid a neighborhood around $\hat{x}$ while converging to the origin. The additional level sets, i.e., $V_1(x) = 0.6$ and $V_2(x) = 0.7$, are used to ensure that the controller selection is well defined and to avoid Zeno behavior. A rigorous analysis of the controller is given in [14].

The ideas presented here are not limited to linear systems. However, while avoidance or stability as isolated problems can be performed through existing methods, guaranteeing avoidance and stability simultaneously for all initial condition is chal-

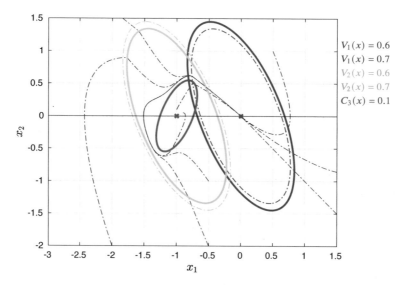

Fig. 5.3 Closed loop solutions using combinations of the control laws u_1, u_2, and u_3 level sets of the of the functions V_1, V_2, and C_3. All solutions asymptotically converge to the origin. Additionally, solutions starting outside the level set $C_3(x) = 0.1$ never enter the sublevel set $C_3(x) \leq 0.1$

lenging. The construction described here relies on the fact that $\hat{x}$ is an induced equilibrium of the linear system (5.10). A similar hybrid controller selection can be performed without this assumption [15].

Chapter 6
Proofs of the Main Results

Abstract In this chapter we prove Theorems 3.13, 3.15 and 4.11. For the proof of Theorem 3.13, the results in [47] are adapted. Theorem 3.15 follows by adapting the arguments in [69].

Keywords Lyapunov methods · Differential inclusions · Stability of nonlinear systems · Instability of nonlinear systems · Stabilization/destabilization of nonlinear systems · Stabilizability and destabilizability

6.1 Proof of Theorem 3.13

In this section we prove Theorem 3.13. The proof follows the ideas in [47] which proves Theorem 3.3 (see [47, Proposition 2.5]), the analogous result to Theorem 3.13 in the context of stability. With a slight variation of the proof of Theorem 3.3, Corollary 4.3 follows. Similarly, Corollary 4.9 follows with minimal changes of the proof of Theorem 3.13 given here.

Before we prove Theorem 3.13, we show that the following intermediate result is true (similar to [47, Lemma 3.1]).

Lemma 6.1 *Consider the differential inclusion (2.1) with $F(x) \subset B_1(0)$ for all $x \in \mathbb{R}^n$ and $F(0) = 0$ satisfying Assumption 2.1. Additionally, assume that there exists a function $\delta \in \mathcal{K}_\infty$ such that for all $\varepsilon > 0$ and for all solutions $\phi \in \mathcal{S}(x_0)$, condition (3.4) is satisfied. Then the following are equivalent.*

(a) For all $x_0 \in \mathbb{R}^n \backslash \{0\}$, for all $\phi \in \mathcal{S}(x_0)$ the limit property (3.5) is satisfied.
(b) For any $r, \varepsilon > 0$ there is a $T > 0$, such that

$$|\phi(t; x_0)| > \varepsilon$$

whenever $|x_0| > r$, $t \geq T$ and $\phi \in \mathcal{S}(x_0)$.

© The Author(s), under exclusive license to Springer Nature Switzerland AG 2021
P. Braun et al., *(In-)Stability of Differential Inclusions*,
SpringerBriefs in Mathematics,
https://doi.org/10.1007/978-3-030-76317-6_6

(c) There exists a family of mappings $T_r : \mathbb{R}_{>0} \to \mathbb{R}_{\geq 0}$, $r > 0$, with the properties

- *for each $r > 0$, T_r is continuous and increasing;*
- *for each $r > 0$, there exists $c_\varepsilon = c_\varepsilon(r) > 0$ such that $T_r(\varepsilon) = 0$ for all $\varepsilon \in (0, \frac{1}{2}c_\varepsilon]$ and $T_r(\varepsilon) > 0$ for all $\varepsilon \in (\frac{1}{2}c_\varepsilon, \infty)$;*
- *for each $r > 0$, it holds that $T_r(\cdot + \frac{1}{2}c_\varepsilon) \in \mathcal{K}$;*
- *$c_\varepsilon : \mathbb{R}_{>0} \to \mathbb{R}_{>0}$ is monotonically increasing and*

$$\lim_{r \searrow 0} c_\varepsilon(r) = 0;$$

- *for each fixed $\varepsilon > 0$, on the domain where $T_r(\varepsilon) \neq 0$, the map $r \mapsto T_r(\varepsilon)$ is (strictly) decreasing*

such that for all $\phi \in S(x_0)$,

$$|\phi(t; x_0)| > \varepsilon + \tfrac{1}{2}c_\varepsilon(r) \quad \text{whenever } |x_0| > r \text{ and } t \geq T_r(\varepsilon + \tfrac{1}{2}c_\varepsilon(r)). \quad (6.1)$$

Since the assumptions of Lemma 6.1 together with item (a) capture strong complete instability according to Definition 3.6, Lemma 6.1 provides alternative presentations of strong complete instability. The properties of item (b) can be understood as *uniform divergence* as a complement to *uniform attractivity*. According to Theorem 2.3, assumption $F(x) \subset B_1(0)$ for all $x \in \mathbb{R}^n$ is not restrictive. In this context the equivalence between (a) and (b) shows that instability properties are uniform in time for autonomous differential inclusions (2.1). The representation of strong complete instability through item (c) is used for the proof of Theorem 3.13.

Proof *"(a) implies (b)":* Fix $r, \varepsilon > 0$ and assume that (b) does not hold. Then there are sequences of initial values $x_k \in \mathbb{R}^n$ with $|x_k| \geq r$, solutions $\phi_k \in S(x_k)$ and times $t_k \to \infty$ as $k \to \infty$ such that $|\phi(t_k; x_0)| \leq \varepsilon$.

Now (3.4) implies that if there is $t^* > 0$ with $|\phi(t^*; x_0)| \geq \delta(2\varepsilon)$ with $\delta(\cdot)$ from (3.4), then $|\phi(t; x_0)| \geq 2\varepsilon > \varepsilon$ for all $t \geq t^*$. From this we can conclude that $|\phi_k(t; x_k)| \leq \delta(2\varepsilon)$ must hold for all $t \in [0, t_k]$. This, in particular, implies that $|x_k| \leq \delta(2\varepsilon)$.

Hence, the x_k are contained in the compact set $\overline{B}_{\delta(2\varepsilon)}(0)$ and thus, by passing to a subsequence if necessary, we can assume that $x_k \to x^*$ as $k \to \infty$ and since $|x_k| \geq r$ for all k it follows that $|x^*| \geq r$. In particular, this implies that $\{x_k | k \in \mathbb{N}\}$ is compact. Since $F(x) \subset B_1(0)$, i.e., $|\dot{\phi}(t; x)| \leq 1$ for almost all $t \geq 0$, for all $x \in \mathbb{R}^n$, and $|x_k| \leq \delta(2\varepsilon)$, all solutions are uniformly bounded on each compact time interval. Hence, we can apply Lemma 7.7 in order to conclude that for each $\tau > 0$ the set $S[0, \tau](\{x_k | k \in \mathbb{N}\})$ of solutions defined for t in $[0, \tau]$ is compact in the space of continuous functions endowed with the ∞-norm. By passing to another subsequence, if necessary, we thus obtain the existence of a solution $\phi^* \in S(x^*)$ such that for all $\tilde{\varepsilon} > 0$ and $\tau > 0$ there is $k_0 \in \mathbb{N}$ with $\sup_{t \in [0, \tau]} |\phi_k(t, x_k) - \phi^*(t, x^*)| < \tilde{\varepsilon}$. This, however, implies that $|\phi^*(t; x^*)| \leq \delta(2\varepsilon) - \tilde{\varepsilon}$ for all $t \in [0, t_k]$ holds for all $\tilde{\varepsilon}$ and all k. From this we can conclude that $|\phi^*(t; x^*)| \leq \delta(2\varepsilon)$ for all $t \geq 0$, which contradicts (a). Hence, (b) must hold.

"(b) implies (a)": This direction follows immediately.

"(b) implies (c)": For $r, \varepsilon > 0$, let

$$A_{r,\varepsilon} = \{T \geq 0 : \forall\, |x_0| \geq r,\ \forall\, \phi \in \mathcal{S}(x_0),\ \forall\, t \geq T,\ |\phi(t; x_0)| \geq \varepsilon\} \subset \mathbb{R}_{\geq 0}. \quad (6.2)$$

From (a) it follows that $A_{r,\varepsilon} \neq \emptyset$ for any $r, \varepsilon > 0$. Moreover,

$$A_{r,\varepsilon_2} \subset A_{r,\varepsilon_1} \quad \text{if } \varepsilon_1 \leq \varepsilon_2 \qquad \text{and} \qquad A_{r_1,\varepsilon} \subset A_{r_2,\varepsilon} \quad \text{if } r_1 \leq r_2.$$

Additionally, for $r, \varepsilon > 0$, we define

$$C_r = \{\varepsilon \geq 0 : \forall\, |x_0| \geq r,\ \forall\, \phi \in \mathcal{S}(x_0),\ \forall\, t \geq 0,\ |\phi(t; x_0)| \geq \varepsilon\} \subset \mathbb{R}_{\geq 0}$$

and it holds that $C_{r_1} \subset C_{r_2}$ if $r_1 \leq r_2$. Based on this definition, we set

$$c_\varepsilon(r) = \sup C_r \qquad\qquad (6.3)$$

and it holds that $c_\varepsilon(r)$ is monotonically increasing with r and $\lim_{r \searrow 0} c_\varepsilon(r) = 0$. Moreover, from (3.4) it follows that $c_\varepsilon(r) \geq \delta^{-1}(r) > 0$ for all $r > 0$.

We return to $A_{r,\varepsilon}$ in (6.2) and define

$$\bar{T}_r(\varepsilon) = \inf A_{r,\varepsilon}.$$

Then, $\bar{T}_r(\varepsilon) < \infty$ for any $r, \varepsilon > 0$ and it holds that

$$\bar{T}_r(\varepsilon_1) \leq \bar{T}_r(\varepsilon_2) \quad \text{if } \varepsilon_1 \leq \varepsilon_2 \qquad \text{and} \qquad \bar{T}_{r_1}(\varepsilon) \geq \bar{T}_{r_2}(\varepsilon) \quad \text{if } r_1 \leq r_2. \qquad (6.4)$$

Moreover, the condition $\bar{T}_r(\varepsilon) = 0$ is satisfied for all $\varepsilon \leq c_\varepsilon(r)$.

For $r, \varepsilon > 0$, we set

$$\hat{T}_r(\varepsilon) = \frac{2}{\varepsilon} \int_{\varepsilon/2}^{\varepsilon} \bar{T}_r(s)\,ds.$$

Since $\bar{T}_r(\cdot)$ is increasing, $\hat{T}_r(\cdot)$ is well defined and locally absolutely continuous. By definition, $\bar{T}_r(\varepsilon) = 0$ for all $\varepsilon \leq c_\varepsilon(r)$ and $\bar{T}_r(\varepsilon) > 0$ for all $\varepsilon > c_\varepsilon(r)$. This implies that $\hat{T}_r(\varepsilon) = 0$ for all $\varepsilon \leq c_\varepsilon(r)$ and $\bar{T}_r(\varepsilon) > 0$ for all $\varepsilon > c_\varepsilon(r)$. Additionally,

$$\hat{T}_r(\varepsilon) \geq \frac{2}{\varepsilon}\bar{T}_r\left(\frac{\varepsilon}{2}\right) \int_{\varepsilon/2}^{\varepsilon} ds = \bar{T}_r\left(\frac{\varepsilon}{2}\right)$$

is satisfied, which again follows from the fact that $\bar{T}_r(\cdot)$ is an increasing function. Furthermore, for $\varepsilon_1 \leq \varepsilon_2$ it holds that

$$\hat{T}_r(\varepsilon_1) = \frac{2}{\varepsilon_1} \int_{\varepsilon_1/2}^{\varepsilon_1} \bar{T}_r(s)ds = \frac{2}{\varepsilon_1} \int_0^{\varepsilon_1/2} \bar{T}_r(s + \tfrac{\varepsilon_1}{2})ds$$

$$\leq \frac{2}{\varepsilon_1} \int_0^{\varepsilon_1/2} \bar{T}_r(s + \tfrac{\varepsilon_2}{2})ds \leq \frac{2}{\varepsilon_2} \int_0^{\varepsilon_2/2} \bar{T}_r(s + \tfrac{\varepsilon_2}{2})ds$$

$$= \hat{T}_r(\varepsilon_2)$$

and both inequalities follow from the fact that $\bar{T}_r(\cdot)$ is increasing. Hence, for all $r > 0$, $\hat{T}_r(\cdot)$ is an increasing (but not necessarily strictly increasing) function. Finally, we select $T_r : \mathbb{R}_{\geq 0} \to \mathbb{R}_{\geq 0}$ as

$$T_r(\varepsilon) = \begin{cases} \hat{T}_r(2\varepsilon), & \text{for } \varepsilon \in [0, \tfrac{1}{2}c_\varepsilon(r)], \\ \hat{T}_r(2\varepsilon) + \frac{2\varepsilon - c_\varepsilon(r)}{r+1}, & \text{for } \varepsilon \in (\tfrac{1}{2}c_\varepsilon(r), \infty), \end{cases} \tag{6.5}$$

which satisfies $T_r(\varepsilon) = 0$ for all $\varepsilon \leq \tfrac{1}{2}c_\varepsilon(r)$ and $T(\cdot + \tfrac{1}{2}c_\varepsilon(r)) \in \mathcal{K}_\infty$.

As a next step we show that $T_{(\cdot)}(\varepsilon)$ is decreasing. Assume to the contrary that there exist $0 < r_1 < r_2$ and $\varepsilon > 0$ such that

$$T_{r_1}(\varepsilon) < T_{r_2}(\varepsilon). \tag{6.6}$$

This implies that

$$\hat{T}_{r_1}(2\varepsilon) + \frac{2\varepsilon - c_\varepsilon(r_1)}{r_1 + 1} < \hat{T}_{r_2}(2\varepsilon) + \frac{2\varepsilon - c_\varepsilon(r_2)}{r_2 + 1}. \tag{6.7}$$

Here, we assume without loss of generality that $2\varepsilon > c_\varepsilon(r_2)$ since otherwise the right-hand side is zero, immediately leading to a contradiction.

Since $\bar{T}_{(\cdot)}(\varepsilon)$ is a decreasing function (see (6.4)), also $\hat{T}_{(\cdot)}(\varepsilon)$ is decreasing. Thus, for (6.7) to hold,

$$\frac{2\varepsilon - c_\varepsilon(r_1)}{r_1 + 1} < \frac{2\varepsilon - c_\varepsilon(r_2)}{r_2 + 1}$$

needs to be satisfied. Furthermore, $r_1 < r_2$ implies the condition $-c_\varepsilon(r_1) < -c_\varepsilon(r_2)$, or equivalently $c_\varepsilon(r_1) > c_\varepsilon(r_2)$. However, this contradicts the fact that $c_\varepsilon(\cdot)$ is increasing that was established after (6.3). Thus, the assumption used in Eq. (6.7) is wrong and $T_{(\cdot)}(\varepsilon)$ is decreasing for all $\varepsilon > 0$.

Similarly it can be shown that $T_{(\cdot)}(\varepsilon)$ is strictly decreasing on the domain where $T_r(\varepsilon) \neq 0$. Assume again to the contrary that there exist $0 < r_1 < r_2$ and $\varepsilon > 0$ such that

$$T_{r_1}(\varepsilon) = \hat{T}_{r_1}(2\varepsilon) + \frac{2\varepsilon - c_\varepsilon(r_1)}{r_1 + 1} = \hat{T}_{r_2}(2\varepsilon) + \frac{2\varepsilon - c_\varepsilon(r_2)}{r_2 + 1} = T_{r_2}(\varepsilon). \tag{6.8}$$

Again we can restrict our attention to $\varepsilon > -c_\varepsilon(r_2)$ since otherwise the right-hand side is zero, implying that the left-hand side is zero. For (6.8) to hold,

$$\frac{2\varepsilon - c_\varepsilon(r_1)}{r_1 + 1} = \frac{2\varepsilon - c_\varepsilon(r_2)}{r_2 + 1} \qquad \Longrightarrow \qquad 2\varepsilon - c_\varepsilon(r_1) < 2\varepsilon - c_\varepsilon(r_2)$$

needs to be satisfied since $r_1 < r_2$. Thus, $c_\varepsilon(r_1) > c_\varepsilon(r_2)$ needs to be satisfied which leads to the same contradiction as before. We can conclude that $T_{(\cdot)}(\varepsilon)$ is strictly decreasing on the domain where $T_r(\varepsilon) \neq 0$.

The only property left to be shown is that T_r defined by (6.5) satisfies (6.1). Let x_0 and t be such that $|x_0| \geq r$ and $t \geq T_r(\varepsilon + \frac{1}{2}c_\varepsilon(r))$. Then for $\varepsilon > 0$,

$$t \geq T_r(\varepsilon + \tfrac{1}{2}c_\varepsilon(r)) \geq \hat{T}_r(2\varepsilon + c_\varepsilon(r)) \geq \bar{T}_r(\varepsilon + \tfrac{1}{2}c_\varepsilon(r)).$$

Hence, by definition of $\bar{T}_r(\varepsilon + \frac{1}{2}c_\varepsilon(r))$, $|\phi(t; x_0)| \geq \varepsilon + \frac{1}{2}c_\varepsilon(r)$, as claimed.
"(c) implies (b)": This direction follows immediately.   □

With Lemma 6.1 we can proceed to prove the main statement.

Proof of Theorem 3.13 *"Definition 3.9 implies Definition 3.6":* Due to the properties of $\kappa \in \mathcal{K}_\infty \mathcal{K}_\infty$, it holds that (3.5) is satisfied for all $\phi \in \mathcal{S}(x_0)$, for all $x_0 \in \mathbb{R}^n \backslash \{0\}$. Define the function

$$\delta(\varepsilon) = \kappa(\varepsilon, 0).$$

Then $\delta \in \mathcal{K}_\infty$ and for all $t \geq 0$, for all $\phi \in \mathcal{S}(x_0)$ it holds that

$$|\phi(t; x_0)| \geq \kappa(|x_0|, t) \geq \kappa(|x_0|, 0) = \delta(|x_0|).$$

Thus, for all $|x_0| \geq \delta(\varepsilon)$ and for all $\phi \in \mathcal{S}(x_0)$ it holds that

$$|\phi(t; x_0)| \geq \delta(|x_0|) \geq \varepsilon,$$

i.e., the first implication is shown.

"Definition 3.6 implies Definition 3.9": Assume that the properties of Definition 3.6 are satisfied. We first show that without loss of generality we can assume that $F(x) \subset B_1(0)$ for all $x \in \mathbb{R}^n$. If this is not the case we can apply Theorem 2.3 and consider the scaled differential inclusion (2.3) with scaling (2.6). Indeed, according to Theorem 2.3, the differential inclusion (2.1) satisfies Definition 3.6 if and only if (2.3) satisfies Definition 3.6 (for the same function $\delta \in \mathcal{K}_\infty$). If we can find $\kappa \in \mathcal{K}_\infty \mathcal{K}_\infty$ such that

$$|\phi_\eta(t; x_0)| \geq \kappa(|x_0|, t) \qquad \forall\, t \geq 0, \ \ \forall\, x_0 \in \mathbb{R}^n, \ \ \forall\, \phi_\eta \in \mathcal{S}_\eta(x_0),$$

then for all $x_0 \in \mathbb{R}^n$ and for all $\phi \in S(x_0)$ there exists a continuous strictly increasing function $\alpha : [0, \infty) \to [0, M)$, $\alpha(0) = 0$, so that

$$|\phi(t; x_0)| = |\phi_\eta(\alpha^{-1}(t); x_0)| \geq \kappa(|x_0|, \alpha^{-1}(t)) \quad \forall t \in [0, M)$$

and $|\phi(t; x_0)| \to \infty$ for $t \to M$. Recall that $S_\eta(\cdot)$ denotes the set of solutions of (2.3), and the property follows from Theorem 2.3.

Moreover, since $\alpha'(s) \leq 1$ and $\alpha(0) = 0$, which follows from (2.6) together with (2.7), it holds that $\alpha(t) \leq t$ (independent of the initial condition and independent of $\phi(\cdot, x_0) \in S(x_0)$) and thus $\kappa(\cdot, \alpha^{-1}(t)) \geq \kappa(\cdot, t)$ for all $t \in [0, M)$. Hence, if we are able to find a $\mathcal{K}_\infty \mathcal{K}_\infty$-function as a lower bound for solutions of the scaled system with scaling (2.6), then the $\mathcal{K}_\infty \mathcal{K}_\infty$-function also satisfies

$$|\phi(t; x_0)| \geq \kappa(|x_0|, t) \quad t \geq 0, \ \forall x_0 \in \mathbb{R}^n, \ \forall \phi \in S(x_0)$$

for the original dynamics (2.1).

Hence, $F(x) \subset B_1(0)$ can indeed be assumed for all $x \in \mathbb{R}^n$, which implies that the assumptions of Lemma 6.1(a) are satisfied. Hence, according to Lemma 6.1(c), for each $r > 0$, there exist functions $Q_r(\cdot) = T_r(\cdot + c_\varepsilon(r)) \in \mathcal{K}$ satisfying the properties in Lemma 6.1.

Moreover, Q_r satisfies $Q_r : \mathbb{R}_{\geq 0} \to [0, M(r))$ where $M(r) \in \mathbb{R}_{>0} \cup \{\infty\}$. (Here, $M(r) = \infty$ if $Q_r \in \mathcal{K}_\infty$ and $M(r) < \infty$ if $Q_r \in \mathcal{K} \setminus \mathcal{K}_\infty$.)

Based on these considerations, $\psi_r(\cdot) = Q_r^{-1}(\cdot) : [0, M(r)) \to \mathbb{R}_{\geq 0}$ is well defined for all $r > 0$. If $M(r) = \infty$, then $\psi_r \in \mathcal{K}_\infty$. If $M(r) < \infty$, then $\psi_r(\cdot)$ satisfies $\psi_r(0) = 0$, $\psi_r(\cdot)$ is strictly increasing and $\psi_r(t) \to \infty$ for $t \to M(r)$. Here, we interpret $\psi_r : \mathbb{R}_{\geq 0} \to \mathbb{R}_{\geq 0} \cup \{\infty\}$ as an extended real valued function and set $\psi_r(t) := \infty$ for $t \geq M(r)$ to ensure that ψ_r is defined for all $t \geq 0$. Then, for all $r > 0$, we select

$$v_r(t) = \min\{\psi_r(t), t\}$$

satisfying $v_r \in \mathcal{K}_\infty$ and $\psi(t) \geq v_r(t)$ for all $t \in \mathbb{R}_{>0}$ (and by construction $v_r(t)$ is finite for all $t \in \mathbb{R}_{\geq 0}$).

Claim 6.2 *For any $|x_0| > r$, for all $\phi \in S(x_0)$ and for all $t \in \mathbb{R}_{\geq 0}$, it holds that $|\phi(t; x_0)| \geq v_r(t) + \frac{1}{2}c_\varepsilon(r)$.*

Proof of Claim 6.2 It follows from the definition of the maps T_r that, for any $r > 0$, $\varepsilon \geq 0$,

$$|x_0| \geq r, \quad t \geq T_r(\varepsilon + \tfrac{1}{2}c_\varepsilon(r)) \quad \Rightarrow \quad |\phi(t; x_0)| \geq \varepsilon + \tfrac{1}{2}c_\varepsilon(r) \quad \forall \phi \in S(x_0).$$

For all $t \in [0, M(r))$ it holds that

$$t = Q_r(Q_r^{-1}(t)) = T_r(\psi_r(t) + \tfrac{1}{2}c_\varepsilon(r))$$

and thus

$$|\phi(t; x_0)| \geq \psi_r(t) + \tfrac{1}{2}c_\varepsilon(r), \qquad \forall \phi \in S(x_0).$$

Since additionally $\psi_r(t) \to \infty$ for $t \to M(r)$, it holds that $|\phi(M(r); x_0)| = \infty$ for all $\phi \in S(x_0)$. Thus for all $t \geq 0$ and for all $\phi \in S(x_0)$ the inequality

$$|\phi(t; x_0)| \geq v_r(t) + \tfrac{1}{2}c_\varepsilon(r)$$

is satisfied. $\qquad\qquad\qquad\qquad\qquad\qquad\qquad\qquad\qquad\qquad\qquad\qquad\qquad\square$

From Lemma 6.1, it follows that $c_\varepsilon(r) > 0$ for all $r > 0$ and we define $c_\varepsilon(0) = 0$ which ensures continuity of c_ε in $r = 0$ since $c_\varepsilon(r) \to 0$ for $r \to 0$, as established in Lemma 6.1(c). Moreover, from (3.4) it follows that

$$|\phi(t; x_0)| \geq \delta^{-1}(r)$$

for all $t \geq 0$, for all $|x_0| \geq r$ and for all $\phi \in S(x_0)$. Thus, based on the definition of $c_\varepsilon(\cdot)$, it holds that $c_\varepsilon(r) \to \infty$ for $r \to \infty$ since $\delta^{-1} \in \mathcal{K}_\infty$. Using [36, Lemma 1], there exists $\rho \in \mathcal{K}_\infty$ such that

$$c_\varepsilon(r) \geq \rho(r), \qquad \forall r \in \mathbb{R}_{\geq 0}.$$

For $r = 0$, we define $\psi_0(t) = 0$ for all $t \in \mathbb{R}_{\geq 0}$ and $v_0(\cdot) = \psi_0(\cdot)$, which implies that

$$|\phi(t; 0)| \geq \psi_0(t) + \tfrac{1}{2}c_\varepsilon(0) = 0 \qquad \forall t \in \mathbb{R}_{>0}, \forall \phi \in S(0).$$

According to Claim 6.2 and the definitions so far, for all $\phi \in S(x_0)$, it holds that

$$|\phi(t; x_0)| \geq \psi_{|x_0|}(t) + \tfrac{1}{2}c_\varepsilon(|x_0|) \geq \inf_{r \in [|x_0|, \infty)} \psi_r(t) + \tfrac{1}{2}c_\varepsilon(r)$$

$$\geq \inf_{r \in [|x_0|, \infty)} v_r(t) + \tfrac{1}{2}\rho(r) \geq \inf_{r \in [|x_0|, \infty)} \left(v_r(t) + \tfrac{1}{4}\rho(r)\right) + \tfrac{1}{4}\rho(|x_0|).$$

We define the function

$$\hat{\psi}(s, t) = \inf_{r \in [s, \infty)} \left(v_r(t) + \tfrac{1}{4}\rho(r)\right) + \tfrac{1}{4}\rho(s).$$

If we can show that $\hat{\psi} \in \mathcal{K}_\infty \mathcal{K}_\infty$ the proof is complete. Since $\rho, v_s \in \mathcal{K}_\infty$ for all $s > 0$ and $v_0 \equiv 0$, it holds that $\hat{\psi}(s, t) \geq 0$ for all $(s, t) \in \mathbb{R}^2_{\geq 0}$ and $\hat{\psi}(0, t) = 0$ for all $t \in \mathbb{R}_{\geq 0}$.

For $t \geq 0$ fixed, it additionally holds that $\hat{\psi}(\cdot, t) \in \mathcal{K}_\infty$ since

$$\inf_{r \in [s, \infty)} \left(v_r(t) + \tfrac{1}{4}\rho(r)\right)$$

is monotonically increasing in s, zero for $s = 0$ and $\tfrac{1}{4}\rho(\cdot) \in \mathcal{K}_\infty$.

For $s > 0$ fixed we consider the function

$$\chi_s(t) = \inf_{r \in [s, \infty)} \left(v_r(t) + \tfrac{1}{4}\rho(r) \right) \tag{6.9}$$

and show that $\chi_s \in \mathcal{K}_\infty$. For $t = 0$ it holds that $\chi_s(0) = 0$. Since $v_r \in \mathcal{K}_\infty$ for all $r > 0$, $\chi_s(\cdot)$ is monotonically increasing. Moreover, $\chi_s(t) \to \infty$ for $t \to \infty$ which follows from the fact that $v_s, \rho \in \mathcal{K}_\infty$. In particular if $r \nrightarrow \infty$ for $t \to \infty$ on the right-hand side of (6.9), then $v_r(t) \to \infty$. Conversely, if $r \to \infty$ then $\rho(r) \to \infty$. This additionally implies that the infimum in (6.9) is attained.

What is left to show is that $\chi_s(t)$ is strictly monotonically increasing. Assume that this is not the case, i.e., there exist $t_1 < t_2$ (and $s > 0$) such that

$$\chi_s(t_1) = \chi_s(t_2). \tag{6.10}$$

Since the infimum is attained, we can write

$$\chi_s(t_1) = v_{s_1}(t_1) + \tfrac{1}{4}\rho(s_1), \qquad \chi_s(t_2) = v_{s_2}(t_2) + \tfrac{1}{4}\rho(s_2)$$

for $s_1, s_2 \in [s, \infty)$.

- Assume that $s_1 = s_2$: In this case $v_{s_1}(t_1) = v_{s_1}(t_2)$ needs to be satisfied for (6.10) to hold, which contradicts the fact that $v_{s_1} \in \mathcal{K}_\infty$.
- Assume that $s_1 < s_2$: In this case we can write again $v_{s_1}(t_1) + \tfrac{1}{4}\rho(s_1) = v_{s_2}(t_2) + \tfrac{1}{4}\rho(s_2)$. However, it holds that

$$v_{s_2}(t_1) + \tfrac{1}{4}\rho(s_2) < v_{s_2}(t_2) + \tfrac{1}{4}\rho(s_2) = v_{s_1}(t_1) + \tfrac{1}{4}\rho(s_1)$$

which contradicts the optimality of s_1 with respect to t_1.
- Assume that $s_2 < s_1$: In this case $v_{s_1}(t_1) + \tfrac{1}{4}\rho(s_1) = v_{s_2}(t_2) + \tfrac{1}{4}\rho(s_2)$ implies that

$$\rho(s_1) > \rho(s_2) \quad \text{and} \quad v_{s_1}(t_1) < v_{s_2}(t_2).$$

Under these assumptions,

$$\chi_s\left(\tfrac{1}{2}(t_1 + t_2)\right) \leq v_{s_2}\left(\tfrac{1}{2}(t_1 + t_2)\right) + \tfrac{1}{4}\rho(s_2) < \chi_s(t_2) = \chi(t_1)$$

which contradicts the fact that χ_s is monotonically increasing.

Continuity of χ_s follows from the continuity of ρ and v_s. In particular, we have shown that $\hat{\psi} \in \mathcal{K}_\infty \mathcal{K}_\infty$, which completes the proof. $\qquad\square$

6.2   Proof of Theorem 3.15

In this section we show that strong $\mathcal{K}_\infty \mathcal{K}_\infty$-instability (Definition 3.9) of the origin of the differential inclusion (2.1) is equivalent to the existence of a smooth Chetaev

function (Definition 3.14). We adapt the proof of [69, Theorem 1] given in the context
of stability and Lyapunov functions to our instability setting.

6.2.1 Preliminary Derivations and Considerations

First note that Theorem 2.3 allows us to rescale the differential inclusion (2.1) without
changing its properties with respect to strong $\mathcal{K}_\infty\mathcal{K}_\infty$-instability of the origin. Thus,
combining the properties derived in Theorem 2.3 and Corollary 2.4 we can assume
without loss of generality that $F(x) \subset \overline{B}_1(0)$ for all $x \in \mathbb{R}^n$ in the derivation of strong
$\mathcal{K}_\infty\mathcal{K}_\infty$-instability properties. Moreover, again due to Corollary 2.4 we can assume
without loss of generality that all solutions of $\dot{x} \in F(x)$ are forward complete.

As a third variation, instead of the differential inclusion (2.1) we consider the
differential inclusion

$$\dot{x} \in F_{\delta(x)}(x), \qquad x(0) \in \mathbb{R}^n \tag{6.11}$$

where the set-valued map $F_{\delta(\cdot)} : \mathbb{R}^n \rightrightarrows \mathbb{R}^n$,

$$F_{\delta(x)}(x) := \overline{\mathrm{conv}}\left(\bigcup_{\xi \in \{x\} + \overline{B}_{\delta(x)}} F(\xi)\right) + \overline{B}_{\delta(x)} \tag{6.12}$$

is defined through an appropriate continuous positive definite function $\delta : \mathbb{R}^n \to$
$[0, 1]$. In this section we give results showing that (6.11) can be used to establish
properties of (2.1). We denote the set-valued map (6.12) by $F_\delta(\cdot) = \overline{\mathrm{conv}}(F(\cdot +$
$\overline{B}_{\delta(\cdot)}) + \overline{B}_{\delta(\cdot)})$. Since $\delta(x) \le 1$ for all $x \in \mathbb{R}^n$ and $F(x) \in \overline{B}_1(0)$, we have that
$F_{\delta(x)}(x) \subset \overline{B}_3(0)$ and forward completeness of solutions of (6.11) follows from
[68, Corollary 1] or [61, Theorem 3.3], as in the proof of Corollary 2.4.

Lemma 6.3 ([69, Lemma 1]) *If the set valued-map F satisfies Assumption 2.1, and*
$F_{\delta(\cdot)}$ *in (6.12) is defined through a continuous function* $\delta : \mathbb{R}^n \to \mathbb{R}_{\ge 0}$, *then the set*
valued-map $F_{\delta(\cdot)}$ *also satisfies the basic condition from Assumption 2.1.*

The following result mirrors [69, Lemma 18] on robust $\mathcal{K}\mathcal{L}$-stability to strong
complete instability.

Lemma 6.4 *Let F satisfy Assumption 2.1 on* $\mathbb{R}^n$ *and suppose that C is a smooth*
Chetaev function according to Definition 3.14 for the differential inclusion (2.1).
Then, there exists a positive function $\delta : \mathbb{R}^n \to \mathbb{R}_{\ge 0}$ *such that the following properties*
are satisfied:

(i) *The set-valued map* $F_\delta(\cdot)$ *satisfies Assumption 2.1 on* $\mathbb{R}^n$.
(ii) *The function* $C_\delta(\cdot) = C(\cdot)^4$ *is a smooth Chetaev function according to Definition*
     3.14 for the differential inclusion (6.11).

The proof follows the same lines as the proof of [69, Lemma 18] with the only difference that the worst case of a decrease in the context of Lyapunov functions is replaced by the worst case of an increase in the context of Chetaev functions. For completeness, we report the proof here.

Proof First note that since C is a smooth Chetaev function (and due to Remark 3.17), in particular the condition

$$\min_{w \in F(x)} \langle \nabla C(x), w \rangle \geq C(x) \tag{6.13}$$

is satisfied.

We start with item (i). We introduce two functions δ_1 and δ_2 as follows. The function δ_1 is defined as

$$\delta_1(x) = \frac{C(x)}{4 \max\{1, |\nabla C(x)|\}} \tag{6.14}$$

and is a positive definite function $\delta_1 : \mathbb{R}^n \to \mathbb{R}_{\geq 0}$ since $C(x) > 0$ for all $x \neq 0$. Additionally, for $x \neq 0$ we define $\delta_2(\cdot)$ through

$$\delta_2(x) = \sup_{\delta \in [0,1]} \delta$$
$$\text{such that} \quad \min_{w \in \overline{\mathrm{conv}}(F(x + \overline{B}_{2\delta}))} \langle \nabla C(x), w \rangle \geq \tfrac{1}{2} C(x). \tag{6.15}$$

For $x = 0$ we set $\delta_2(0) = 0$ and show as a next step that δ_2 is well-defined.

Since F is upper semicontinuous, for all $x \in \mathbb{R}^n \backslash \{0\}$ there exists $\bar{\delta}(x) > 0$ such that

$$F(x + \overline{B}_{2\bar{\delta}(x)}) \subset F(x) + \overline{B}_{2\delta_1(x)}.$$

Then, using the convexity of $F(x)$ and [30, Sect. 5, Lemma 9], it holds that

$$\overline{\mathrm{conv}}(F(x + \overline{B}_{2\bar{\delta}(x)})) \subset \overline{\mathrm{conv}}(F(x) + \overline{B}_{2\delta_1(x)})$$
$$= F(x) + \overline{B}_{2\delta_1(x)} \tag{6.16}$$

and it follows from the definition of $\delta_1(x)$ in (6.14) that

$$\min_{w \in \overline{\mathrm{conv}}(F(x + \overline{B}_{2\delta}))} \langle \nabla C(x), w \rangle \geq \min_{w \in F(x) + \overline{B}_{2\delta_1(x)}} \langle \nabla C(x), w \rangle \geq \tfrac{1}{2} C(x) \tag{6.17}$$

is satisfied with $\delta = \min\{1, \bar{\delta}(x)\}$.

We claim that for each compact subset $\mathcal{A} \subset \mathbb{R}^n \backslash \{0\}$ we have $\inf_{x \in \mathcal{A}} \delta_2(x) > 0$. Suppose to the contrary that this is not the case. Then there exists a sequence $(x_k)_{k \in \mathbb{N}}$, with $x_k \in \mathcal{A}$ for all $k \in \mathbb{N}$ and $\lim_{k \to \infty} x_k = x^\star \in \mathcal{A}$, such that

$$\min_{w \in \overline{\text{conv}}(F(x_k + \overline{B}_{\frac{1}{k}}))} \langle \nabla C(x_k), w \rangle < \tfrac{1}{2} C(x_k). \tag{6.18}$$

Using the upper semicontinuity of F and the convexity of $F(x^\star)$, as in (6.16), we have for k sufficiently large,

$$\overline{\text{conv}}(F(x_k + \overline{B}_{\frac{1}{k}})) \subset F(x^\star) + \overline{B}_{\delta_1(x^\star)} \tag{6.19}$$

and thus

$$\tfrac{1}{2} C(x_k) > \min_{w \in \overline{\text{conv}}(F(x_k + \overline{B}_{\frac{1}{k}}))} \langle \nabla C(x_k), w \rangle$$
$$\geq \min_{w \in F(x^\star) + \overline{B}_{\delta_1(x^\star)}} \langle \nabla C(x_k), w \rangle. \tag{6.20}$$

Since (6.14) gives

$$\min_{w \in F(x^\star) + \overline{B}_{\delta_1(x^\star)}} \langle \nabla C(x_k), w \rangle \geq \tfrac{3}{4} C(x^\star) \tag{6.21}$$

the continuity of C and the estimates (6.20) and (6.21) thus provide a contradiction to (6.18) and we can conclude that $\delta_2(x) > 0$ for all $x \neq 0$.

We define the function

$$\Delta(x) = \min\{\delta_1(x), \delta_2(x), 1\}, \tag{6.22}$$

which provides the estimate

$$\min_{w \in \overline{\text{conv}}(F(x + \overline{B}_{\Delta(x)}) + \overline{B}_{\Delta(x)})} \langle \nabla C(x), w \rangle \geq \tfrac{1}{2} C(x) \geq \tfrac{1}{4} C(x), \tag{6.23}$$

i.e.,

$$\min_{w \in \overline{\text{conv}}(F(x + \overline{B}_{\Delta(x)}) + \overline{B}_{\Delta(x)})} \langle \nabla (C(x)^4), w \rangle \geq C(x)^4. \tag{6.24}$$

Now, we can define the function δ as

$$\delta(x) = \inf_{\xi \in \mathbb{R}^n} (\Delta(\xi) + |x - \xi|),$$

the so-called inf-convolution. The function δ is Lipschitz continuous and satisfies $\delta(x) \leq \Delta(x) \leq 1$ and $\Delta(x) > 0$ implies $\delta(x) > 0$. Thus, Lemma 6.3 ensures that Assumption 2.1 is satisfied for $F_{\delta(\cdot)}(\cdot)$, which implies that δ satisfies the conditions of Lemma 6.4 item (i).

With item (i) the proof of item (ii) follows immediately. Since $\Delta(x) \geq \delta(x)$, inequality (6.24) implies

$$\min_{w \in F_\delta(x)} \langle \nabla(C(x)^4), w \rangle \geq C(x)^4.$$

Additionally, if

$$\alpha_1(|x|) \leq C(x) \leq \alpha_2(|x|)$$

for $\alpha_1, \alpha_2 \in \mathcal{K}_\infty$ then

$$\alpha_1(|x|)^4 \leq C(x)^4 \leq \alpha_2(|x|)^4$$

and $\alpha_1(\cdot)^4, \alpha_2(\cdot)^4 \in \mathcal{K}_\infty$. Thus $C(\cdot)^4$ is a smooth Chetaev function according to Definition 3.14 with respect to the differential inclusion (6.11).   □

To distinguish between solutions of the differential inclusion (2.1) and (6.11) we use $\phi(\cdot; x_0)$ and $\mathcal{S}(x_0)$, and $\phi_\delta(\cdot; x_0)$ and $\mathcal{S}_\delta(x_0)$ in the following to denote a solution and the set of solutions of (2.1) and (6.11), respectively.

Definition 6.5 The equilibrium $0 \in \mathbb{R}^n$ is robustly strongly completely unstable with respect to the differential inclusion (2.1) if there exist $\kappa \in \mathcal{K}_\infty \mathcal{K}_\infty$ and a positive definite function $\delta : \mathbb{R}^n \to \mathbb{R}_{\geq 0}$ such that, for all $x_0 \in \mathbb{R}^n$ every solution $\phi_\delta \in \mathcal{S}_\delta(x_0)$ of (6.11) satisfies

$$|\phi_\delta(t; x_0)| \geq \kappa(|x_0|, t), \quad \forall\, t \in \mathbb{R}_{\geq 0}. \tag{6.25}$$

Since for $\delta : \mathbb{R}^n \to \mathbb{R}_{\geq 0}$ the set of solutions $\mathcal{S}(x_0)$ is contained in $\mathcal{S}_\delta(x_0)$, i.e., $\mathcal{S}(x_0) \subset \mathcal{S}_\delta(x_0)$ for all $x_0 \in \mathbb{R}^n$, robust strong complete instability implies strong complete instability. More surprisingly, as we have shown, also the other direction is true (cf. also [69, Theorem 1] for an analogous statement in the case of stability).

6.2.2 Smooth Chetaev Functions Imply Strong Complete Instability

In this section we show that the existence of a smooth Chetaev function implies robust strong complete instability following [69, Sect. 5.1.1.]. Different to [69, Sect. 5.1.1.] we do not assume forward completeness since, as argued, it can be obtained without loss of generality. Apart from forward completeness, the instability result follows with minimal changes to the arguments used for the stability result.

Proposition 6.6 *Let F satisfy Assumption 2.1 on $\mathbb{R}^n$. Let C be a smooth Chetaev function according to Definition 3.14 for the differential inclusion (2.1). Then the origin of the differential inclusion (2.1) is robustly strongly completely unstable according to Definition 6.5.*

From this result the first implication of Theorem 3.15 follows.

Proof Let C be a smooth Chetaev function according to Definition 3.14 and let δ satisfy the assumptions of Lemma 6.4. Additionally, we assume without loss of generality that C satisfies (3.14) (since otherwise C can be rescaled according to Remark 3.17).

According to Lemma 6.3, the differential inclusion (6.11) satisfies Assumption 2.1. Lemma 6.4 provides $\alpha_1, \alpha_2 \in \mathcal{K}_\infty$ such that

$$\alpha_1(|\phi_\delta(t; x_0)|) \leq C(\phi_\delta(t; x_0))^4 \leq \alpha_2(|\phi_\delta(t; x_0)|) \tag{6.26}$$

and

$$\overbrace{C(\phi_\delta(t; x_0))^4}^{\cdot} \geq C(\phi_\delta(t; x_0))^4 \tag{6.27}$$

for all solutions of the differential inclusion (6.11) and all $t \geq 0$ such that $\phi_\delta(t; x)$ is finite. For $|\phi(t; x)| = \infty$, the inequalities (6.26) and (6.27) are satisfied by assumption. Thus, from (6.27) it follows that

$$C(\phi_\delta(t; x_0))^4 \geq C(x_0)^4 e^t \tag{6.28}$$

for all solutions, for all $t \geq 0$ such that $|\phi_\delta(t; x)| < \infty$ and for all initial states $x \in \mathbb{R}^n$. Inequality (6.28) follows from (6.27) by dividing (6.27) through the right-hand side, integrating both sides from zero to t and rearranging the terms.

Then (6.26) implies that

$$|\phi_\delta(t; x_0)| \geq \alpha_2^{-1}(\alpha_1(x_0)e^t) \tag{6.29}$$

for all $t \geq 0$ such that $|\phi_\delta(t; x_0)| < \infty$.

Since the right-hand side in (6.29) is finite for all $t \in \mathbb{R}_{\geq 0}$ and for all $x_0 \in \mathbb{R}^n$, (6.29) holds for all $t \geq 0$ even if the restriction $|\phi_\delta(t; x_0)| < \infty$ is dropped. Hence, the origin of the differential inclusion (6.11) is strongly $\mathcal{K}_\infty\mathcal{K}_\infty$-unstable (see Definition 3.9) and the origin of the original differential inclusion (2.1) is robustly strongly completely unstable (see Definition 6.5). □

6.2.3 Strong Complete Instability Implies the Existence of a Smooth Chetaev Function

In this section we prove that strong $\mathcal{K}_\infty\mathcal{K}_\infty$-instability implies the existence of a smooth Chetaev function according to Definition 3.14. We adapt the proof given in [69, Sect. 5.1.2] showing that robust $\mathcal{K}\mathcal{L}$-stability implies the existence of a smooth Lyapunov function.

Theorem 6.7 *Consider the differential inclusion (2.1), let Assumption 2.1 be satis-fied and assume that the origin of (2.1) is robustly strongly $\mathcal{K}_\infty\mathcal{K}_\infty$-unstable accord-ing to Definition 6.5. Then there exists a smooth Chetaev function according to Definition 3.14.*

Proof The proof is split in several parts, following [69, Sect. 5.1.2].
Step 1: Preliminaries: Following the arguments in Sect. 6.2.1, we assume without loss of generality that all solutions of $\dot{x} \in F(x)$ and all solutions of $\dot{x} \in F_{\delta(x)}(x)$ are finite for all $t \in [0, \infty)$. Let $\delta : \mathbb{R}^n \to [0, 1]$ be the continuous positive definite function for robust strong complete instability in Definition 6.5. According to Lemma 7.9 there exists a set-valued map $F_L : \mathbb{R}^n \rightrightarrows \mathbb{R}^n$ satisfying Assumptions 2.1, 2.2 and

$$F(x) \subset F_L(x) \subset F_\delta(x) \quad \forall \, x \in \mathbb{R}^n.$$

Since $\dot{x} \in F_\delta(x)$ is robustly strongly completely unstable by assumption, the dif-ferential inclusion $\dot{x} \in F_L(x)$ is robustly strongly completely unstable. The set of solutions of the differential inclusion $\dot{x} \in F_L(x)$ is denoted by $\mathcal{S}_L(x)$ and all solu-tions are forward complete since all solutions of $\dot{x} \in F_\delta(x)$ are forward complete and $\mathcal{S}_L(x) \subset \mathcal{S}_\delta(x)$.

Step 2: Construction of C_1 and its basic properties: Let $\kappa \in \mathcal{K}_\infty\mathcal{K}_\infty$ be such that for each $x_0 \in \mathbb{R}^n$ every solution $\phi \in \mathcal{S}_L(x_0)$ satisfies

$$|\phi(t; x_0)| \geq \kappa(|x_0|, t) \quad \forall \, t \in \mathbb{R}_{\geq 0}.$$

According to Lemma 7.2 for $\lambda = 2$, there exist $\alpha_1, \alpha_2 \in \mathcal{K}_\infty$ such that

$$\alpha_2(|\phi(t; x_0)|) \geq \alpha_2(\kappa(|x_0|, t)) \geq \alpha_1(|x_0|)e^{2t} \tag{6.30}$$

for all $x_0 \in \mathbb{R}^n$, for all $\phi \in \mathcal{S}_L(x_0)$ and for all $t \in \mathbb{R}_{\geq 0}$. For each $x_0 \in \mathbb{R}^n$ we define

$$C_1(x_0) = \inf_{t \geq 0; \, \phi \in \mathcal{S}_L(x_0)} \alpha_2(|\phi(t; x_0)|)e^{-t}. \tag{6.31}$$

We claim that the function $C_1 : \mathbb{R}^n \to \mathbb{R}$ has the following properties.

Claim 6.8 *The function $C_1 : \mathbb{R}^n \to \mathbb{R}$ defined in (6.31) satisfies the following prop-erties:*

(i) C_1 is positive definite, i.e., $C_1(x) = 0$ if and only if $x = 0$.
(ii) C_1 satisfies the bounds

$$\alpha_1(|x|) \leq C_1(x) \leq \alpha_2(|x|), \tag{6.32}$$

(iii) If C_1 is continuous and locally Lipschitz continuous on $\mathbb{R}^n \backslash \{0\}$, then it holds that

$$\min_{w \in F_L(x)} \langle \nabla C_1(x), w \rangle \geq C_1(x) \tag{6.33}$$

for almost all $x \in \mathbb{R}^n$.

Proof of Claim 6.8 (i) The first item follows from the definition of C_1 in (6.31) combined with the estimate (6.30).

(ii) Based on the definition of C_1 in (6.31) and the estimate (6.30) it holds that

$$C_1(x) = \inf_{t\geq 0;\ \phi\in S_L(x)} \alpha_2(|\phi(t;x)|)e^{-t} \geq \inf_{t\geq 0, \phi\in S_L(x)} \alpha_1(|x|)e^{2t}e^{-t} \geq \alpha_1(|x|)$$

and which provides the lower bound. The upper bound in (6.32) is obtained through

$$C_1(x) = \inf_{t\geq 0;\ \phi\in S_L(x)} \alpha_2(|\phi(t;x)|)e^{-t}$$

$$\leq \inf_{\phi\in S_L(x)} \alpha_2(|\phi(t;x)|)e^{-t}\Big|_{t=0}$$

$$= \alpha_2(|\phi(0,x)|) = \alpha_2(|x|).$$

(iii) To establish (6.33) we assume that C_1 is continuous and locally Lipschitz continuous on $\mathbb{R}^n\setminus\{0\}$ and first show that the inequality

$$C_1(\phi(t;x)) \geq C_1(x)e^t \tag{6.34}$$

is satisfied for all $x \in \mathbb{R}^n$, for all $\phi \in S_L(x)$ and for all $t \in \mathbb{R}_{\geq 0}$. It holds that

$$C_1(\phi(t;x)) = \inf_{\tau\geq 0,\ \varphi\in S_L(\phi(t;x))} \alpha_2(|\varphi(\tau,\phi(t;x))|)e^{-\tau}$$

$$\geq \inf_{\tau\geq t,\ \varphi\in S_L(x)} \alpha_2(|\varphi(\tau,x)|)e^{-(\tau-t)}$$

$$\geq \inf_{\tau\geq 0,\ \varphi\in S_L(x)} \alpha_2(|\varphi(\tau,x)|)e^{-\tau}e^t$$

$$= C_1(x)e^t.$$

Since F_L is locally Lipschitz on $\mathbb{R}^n\setminus\{0\}$, from Lemma 7.12, for each $w \in F_L(x)$ there exists $\phi \in S_L(x)$ such that

$$\phi(t;x) = x + t(w + r(t)), \qquad \forall\, t \in [0,T),$$

for some $T > 0$ and some continuous function r satisfying $\lim_{t\searrow 0} r(t) = 0$. Then from (6.34) we have

$$\tfrac{1}{t}(C_1(x + t(w + r(t))) - C_1(x)) \geq C_1(x)\tfrac{e^t-1}{t}.$$

for $t > 0$ sufficiently small. With the definition of the (upper right) Dini derivative in (2.19a), for $x \neq 0$, we have

$$D^+C_1(x; w) \geq \limsup_{t \searrow 0} \tfrac{1}{t}(C_1(x + t(w + r(t))) - C_1(x)) \geq C_1(x).$$

Here, $D^+C_1(x; w)$ is finite according to the discussion in Sect. 2.3 since C_1 is locally Lipschitz continuous by assumption. Moreover, $\frac{e^t-1}{t} \to 1$ for $t \to 0$ is used here and which can be obtained from L'Hôpital's rule. Since $w \in F_L(x)$ was arbitrary, for each $x \in \mathbb{R}^n \backslash \{0\}$ it holds that

$$\inf_{w \in F_L(x)} D^+C_1(x; w) \geq C_1(x). \tag{6.35}$$

Thus, the assertion follows from (6.35) and the fact that Lipschitz continuous functions are continuously differentiable almost everywhere as discussed in Sect. 2.3.

$\square$

Since Claim 6.8(iii) relies on C_1 being continuous and locally Lipschitz on $\mathbb{R}^n \backslash \{0\}$ we establish these properties next.

Step 3: Continuity of C_1: From the definition of C_1 in (6.31) it follows that

$$0 \leq C_1(x) = \inf_{t \geq 0;\ \phi \in S_L(x)} \alpha_2(|\phi(t; x)|)e^{-t} \leq \inf_{\phi \in S_L(x)} \alpha_2(|\phi(0; x)|)e^0 = \alpha_2(|x|)e. \tag{6.36}$$

Thus, $C_1(x) \to C_1(0) = 0$ for $|x| \to 0$ and we can conclude that C_1 is continuous at $x = 0$.

Claim 6.9 *Consider*

$$T(x) = \ln\left(\frac{C_1(x)}{\alpha_1(|x|)}\right) + 1 \tag{6.37}$$

for $x \in \mathbb{R}^n \backslash \{0\}$. Then, T is well defined and $T : \mathbb{R}^n \backslash \{0\} \to [1, \infty)$. Moreover, there exists a $\psi \in S_L(x)$ such that

$$C_1(x) = \min_{t \in [0, T(x)]} \alpha_2(|\psi(t; x)|)e^{-t}. \tag{6.38}$$

Proof of Claim 6.9 Since $C_1(x) > 0$ for all $x \neq 0$ by the definition in (6.31), and due to the property in Claim 6.8(ii), T is well defined and $T : \mathbb{R}^n \backslash \{0\} \to [1, \infty)$.

The definition of C_1 in (6.31) implies that

$$\inf_{t \in [0, T(x)],\ \phi \in S_L(x)} \alpha_2(|\phi(t; x)|)e^{-t} \geq C_1(x)$$

where on the left-hand side, $t \in \mathbb{R}_{\geq 0}$ is replaced by $t \in [0, T(x)]$. With (6.30) additionally the estimates

$$C_1(x) = \min \left\{ \inf_{\substack{t \in [0,T(x)] \\ \phi \in S_L(x)}} \alpha_2(|\phi(t;x)|)e^{-t}, \inf_{\substack{t \geq T(x) \\ \phi \in S_L(x)}} \alpha_2(|\phi(t;x)|)e^{-t} \right\}$$

$$\geq \min \left\{ \inf_{\substack{t \in [0,T(x)] \\ \phi \in S_L(x)}} \alpha_2(|\phi(t;x)|)e^{-t}, \inf_{t \geq T(x)} \alpha_1(|x|)e^{2t}e^{-t} \right\}$$

$$\geq \min \left\{ \inf_{\substack{t \in [0,T(x)] \\ \phi \in S_L(x)}} \alpha_2(|\phi(t;x)|)e^{-t}, \alpha_1(|x|)e^{T(x)} \right\}$$

$$= \min \left\{ \inf_{\substack{t \in [0,T(x)] \\ \phi \in S_L(x)}} \alpha_2(|\phi(t;x)|)e^{-t}, C_1(x)e \right\} \qquad (6.39)$$

are satisfied and the last equation follows from the definition of T in (6.37) since

$$e^{T(x)} = e^{\ln\left(\frac{C_1(x)}{\alpha_1(|x|)}\right)+1} = \frac{C_1(x)}{\alpha_1(|x|)}e.$$

Thus, it holds that

$$C_1(x) = \inf_{\substack{t \in [0,T(x)] \\ \phi \in S_L(x)}} \alpha_2(|\phi(t;x)|)e^{-t} \qquad (6.40)$$

$$= \inf_{\phi \in S_L(x)} \min_{t \in [0,T(x)]} \alpha_2(|\phi(t;x)|)e^{-t}. \qquad (6.41)$$

and (6.40) follows from (6.39) together with the fact that $C_1(x) \leq C_1(x)e$ for all $x \in \mathbb{R}^n$ and (6.41) follows from continuity of $\alpha_2(|\phi(\cdot;x)|)$. Finally, let $(\phi_k)_{k \in \mathbb{N}} \subset S_L(x)$ be a minimizing sequence, i.e.,

$$C_1(x) = \lim_{k \to \infty} \min_{t \in [0,T(x)]} \alpha_2(|\phi_k(t;x)|)e^{-t}.$$

According to Lemma 7.7, a subsequence of $(\phi_k)_{k \in \mathbb{N}} \subset S_L(x)$ converges uniformly on $[0, T(x)]$ to some $\psi \in S_L(x)$. Then, continuity of $\alpha_2(|\cdot|)$ implies that

$$C_1(x) = \min_{t \in [0,T(x)]} \alpha_2(|\psi(t;x)|)e^{-t}$$

and which completes the proof. □

As a next step we show that C_1 is lower semicontinuous.

Claim 6.10 *The function C_1 defined in (6.31) is lower semicontinuous on $\mathbb{R}^n$, i.e.,*

$$\liminf_{x_k \to x} C_1(x_k) \geq C_1(x) \qquad \forall x \in \mathbb{R}^n.$$

Proof of Claim 6.10 We prove the statement by contradiction. Suppose there exists an $x \in \mathbb{R}^n$ and a sequence $(x_k)_{k \in \mathbb{N}} \subset \mathbb{R}^n$, $x_k \to x$, for $k \to \infty$ such that

$$0 \leq \liminf_{k \to \infty} C_1(x_k) < C_1(x). \tag{6.42}$$

Without loss of generality, we can assume that, for all k and some $\nu > 0$ it holds that

$$C_1(x_k) \leq \nu < C_1(x) \tag{6.43}$$

and we define $\tau = \sup_{k \in \mathbb{N}} T(x_k)$. Condition (6.43) together with continuity of $\alpha(|\cdot|)$ and the definition of $T(\cdot)$ imply that $\tau < \infty$ is satisfied. Let $\psi_k \in S_L(x_k)$ be defined through Eq. (6.38). Then it holds that

$$C_1(x_k) = \min_{t \in [0, T(x_k)]} \alpha_2(|\psi_k(t; x_k)|)e^{-t} = \min_{t \in [0, \tau]} \alpha_2(|\psi_k(t; x_k)|)e^{-t}.$$

For each $\varepsilon > 0$ (and $\varepsilon < \nu$), Lemma 7.8 with the triple $(\tau, \varepsilon, \{x\})$ guarantees the existence of $k_\varepsilon \in \mathbb{N}$ so that for all $k \geq k_\varepsilon$, we can find $\varphi_k \in S_L(x)$ with the property

$$C_1(x_k) = \min_{t \in [0, \tau]} \alpha_2(|\psi_k(t; x_k)|)e^{-t} \geq \min_{t \in [0, \tau]} \alpha_2(|\varphi_k(t; x)|)e^{-t} - \varepsilon \geq C_1(x) - \varepsilon.$$

This implies that

$$\liminf_{k \to \infty} C_1(x_k) \geq C_1(x),$$

a contradiction to assumption (6.42), and thus it establishes lower semicontinuity of C_1 for $x \neq 0$. For $x = 0$ we have already shown continuity of C_1 in (6.36). $\square$

     As a next step we show that the function C_1 is upper semicontinuous. Then, continuity of C_1 follows from the lower and the upper semicontinuity. We prove upper semicontinuity through the two following statements.

Claim 6.11 *Let $x \neq 0$ and let ψ be defined according to (6.38). There exists $\hat{T}(x) \in [0, T(x)]$ such that*

$$C_1(\psi(t; x)) \leq C_1(x)e^{T(x)} \quad \forall t \in [0, \hat{T}(x)], \tag{6.44}$$

and

$$C_1(x) = \min_{t \in [0, \hat{T}(x)]} \alpha_2(|\psi(t; x)|)e^{-t}. \tag{6.45}$$

Claim 6.12 *The function C_1 is upper semicontinuous on $\mathbb{R}^n$, i.e.,*

$$\liminf_{x_k \to x} C_1(x_k) \leq C_1(x) \quad \forall x \in \mathbb{R}^n.$$

Proof of Claim 6.11 First recall that (6.34) implies

$$C_1(\psi(T(x), x)) \geq C_1(x)e^{T(x)}.$$

Thus, the set

$$\mathcal{T} = \{t \in [0, T(x)] : C_1(\psi(t; x)) \geq C_1(x)e^{T(x)}\}$$

is nonempty and we can define

$$\hat{T}(x) = \inf\{t \in \mathcal{T}\}.$$

Either $\hat{T}(x) = 0$ and (6.44) holds with $C_1(\psi(0, x)) = C_1(x) \leq C_1(x)e^{T(x)}$, or $\hat{T}(x) > 0$ and (6.44) holds for all $t \in [0, \hat{T}(x))$.

Additionally, with the lower semicontinuity of C_1 at $\psi(\hat{T}(x); x)$, we have

$$C_1(\psi(\hat{T}(x); x)) \leq \liminf_{z \to \psi(\hat{T}(x); x)} C_1(z) \leq \liminf_{\substack{t \to \hat{T}(x) \\ t < \hat{T}(x)}} C_1(\psi(t; x)) \leq C_1(x)e^{T(x)}$$

which shows that (6.44) also holds for $t = \hat{T}(x)$.

To complete the proof we focus on Eq. (6.45). If $\hat{T}(x) = T(x)$ then there is nothing to prove. For $\hat{T}(x) \neq T(x)$ let $(t_n)_{n \in \mathbb{N}} \subset [0, T(x)) \cap \mathcal{T}$ be a nonincreasing sequence with $t_n \to \hat{T}(x)$ for $n \to \infty$. It holds that

$$C_1(x) = \min\left\{ \min_{t \in [0, t_n]} \alpha_2(|\psi(t; x)|)e^{-t}, \min_{t \in [t_n, T(x)]} \alpha_2(|\psi(t; x)|)e^{-t} \right\}$$

$$\geq \min\left\{ \min_{t \in [0, t_n]} \alpha_2(|\psi(t; x)|)e^{-t}, \inf_{\substack{t \in \mathbb{R}_{\geq 0} \\ \phi \in S_L(\psi(t_n; x))}} \alpha_2(|\phi(t; \psi(t_n; x))|)e^{-t}e^{-t_n} \right\}$$

$$\geq \min\left\{ \min_{t \in [0, t_n]} \alpha_2(|\psi(t; x)|)e^{-t}, C_1(\psi(t_n; x))e^{-t_n} \right\}. \tag{6.46}$$

Moreover, since $t_n \in \mathcal{T}$ and $t_n \leq T(x)$, the inequality

$$C_1(\psi(t_n; x))e^{-t_n} \geq C_1(x)e^{T(x)-t_n} \geq C_1(x)$$

is satisfied. Hence, from (6.46) together with continuity of the functions $\alpha_2(|\cdot|)$ and $\psi(\cdot; x)$, and the fact that $t_n \to \hat{T}(x)$ for $n \to \infty$ establishes (6.45). □

Proof of Claim 6.12 Let $x \in \mathbb{R}^n \backslash \{0\}$, let $\hat{T}(x)$ come from Claim 6.11 and $\psi(t; x)$ be a solution such that (6.38) is satisfied. Recall that the function F_L is locally Lipschitz on $\mathbb{R}^n \backslash \{0\}$ by assumption. Since $|\psi(t; x)| \geq \kappa(|x|, t)$, for all $t \in \mathbb{R}_{\geq 0}$, it holds that $\psi(t; x) \neq 0$ for all $t \geq 0$. With the definitions

$$T = \hat{T}(x), \qquad \mathcal{A} = \{z \in \mathbb{R}^n \mid \exists t \in [0, \hat{T}(x)] : z = \psi(t; x)\}$$

and since $\alpha_2(|\cdot|)$ is locally Lipschitz continuous (which can be assumed without loss of generality), we can apply Lemma 7.10 to conclude the existence of $\delta_x > 0$ and L_x such that for any $|v| \leq \delta_x$, there exists a solution $\phi \in \mathcal{S}_L(x + v)$ such that

$$\max_{t \in [0, \hat{T}(x)]} |\alpha_2(|\psi(t; x)|) - \alpha_2(|\phi(t; x + v)|)| \leq L_x |v|.$$

This yields

$$C_1(x) = \min_{t \in [0, \hat{T}(x)]} \alpha_2(|\phi(t; x)|) e^{-t}$$

$$\geq \min_{t \in [0, \hat{T}(x)]} \alpha_2(|\phi(t; x + v)|) e^{-t} - \max_{t \in [0, \hat{T}(x)]} |\alpha_2(|\phi(t; x + v)|) - \alpha_2(|\psi(t; x)|)| \, e^{-t}$$

$$\geq \inf_{t \geq 0} \alpha_2(|\phi(t; x + v)|) e^{-t} - e^0 L_x |v| \geq C_1(x + v) - e L_x |v| \qquad (6.47)$$

from which upper semicontinuity can be concluded, i.e.,

$$\limsup_{v \to 0} (C_1(x + v) - e L_x |v|) = \limsup_{v \to 0} C_1(x + v) \leq \limsup_{v \to 0} C_1(x) = C_1(x).$$

$\square$

From lower semicontinuity (see Claim 6.10), upper semicontinuity (see Claim 6.12) and continuity of C_1 at $x = 0$ established in (6.36) it follows that C_1 is continuous on $\mathbb{R}^n$. As a next step we show that C_1 is Lipschitz continuous on $\mathbb{R}^n \backslash \{0\}$. *Step 4: Lipschitz continuity of C_1 on $\mathbb{R}^n \backslash \{0\}$:* To show Lipschitz continuity of C_1, we apply Lemma 7.13, i.e., we show the existence of $M > 0$ such that

$$D_+ C_1(x; v) \leq M |v| \qquad \forall x \in \mathcal{U}, \ \forall v \in \mathbb{R}^n, \qquad (6.48)$$

where $\mathcal{U} \subset \mathbb{R}^n$ is open and convex.

Claim 6.13 *For all $x \in \mathbb{R}^n \backslash \{0\}$ there exists a convex neighborhood $\mathcal{U} \subset \mathbb{R}^n$ and a constant $M > 0$ such that (6.48) is satisfied.*

Proof of Claim 6.13 Since $C_1(\cdot)$ and $T(\cdot)$ are continuous functions on $\mathbb{R}^n \backslash \{0\}$, there exists a compact subset $\mathcal{A}_0 \subset \mathbb{R}^n \backslash \{0\}$ which contains a neighborhood of x and is such that for all $z \in \mathcal{A}_0$, we have

$$T(z) \leq 2T(x) \quad \text{and} \quad C_1(z) e^{T(z)} \leq 2 C_1(x) e^{T(x)}. \qquad (6.49)$$

These inequalities together with (6.44) imply that for all $z \in \mathcal{A}_0$ and for all $t \in [0, \hat{T}(z)]$, the inequalities

$$C_1(\psi(t; z)) \leq C_1(z) e^{T(z)} \leq 2 C_1(x) e^{T(x)} \qquad (6.50)$$

are satisfied.

On the other hand, from Lemma 7.7, we know that $\mathcal{R}_{\leq 2T(x)}(\mathcal{A}_0)$ is a compact subset of $\mathbb{R}^n$. Also, since C_1 is continuous on $\mathbb{R}^n$, the following set is a compact subset of $\mathbb{R}^n$:

$$\mathcal{A} = \mathcal{R}_{\leq 2T(x)}(\mathcal{A}_0) \cap \{z \in \mathbb{R}^n \mid C_1(z) \leq 2C_1(x)e^{T(x)}\}. \tag{6.51}$$

From (6.49) and (6.45), we have that for all $z \in \mathcal{A}_0$, and for all $t \in [0, \hat{T}(z)]$, $\psi(t; z) \in \mathcal{A}$ (and ψ satisfies (6.38)). Since $\alpha_2(|\cdot|)$ is locally Lipschitz continuous, we apply Lemma 7.10 with $T = 2T(x)$ and $\mathcal{A}$ defined in (6.51), to conclude the existence of $\delta > 0$ and L such that for all $\bar{v} \in \mathbb{R}^n$ with $|\bar{v}| \leq \delta$, and $z \in \mathcal{A}_0$, we have (following the same lines as for establishing (6.47)),

$$C_1(z) \geq C_1(z + \bar{v}) - Le|\bar{v}|.$$

From $C_1(z + \bar{v}) - C_1(z) \leq Le|\bar{v}|$ for all $|\bar{v}| \leq \delta$ it follows that

$$D_+C_1(x; w) = \liminf_{v \to w;\ \varepsilon \searrow 0} \frac{C_1(x + tv) - C_1(x)}{\varepsilon} \leq \liminf_{v \to w;\ \varepsilon \searrow 0} \frac{Le|\varepsilon w|}{\varepsilon} = (Le)|w|$$

which establishes (6.48) and thus completes the proof. $\square$

We can conclude that C_1 is Lipschitz continuous on $\mathbb{R}^n \backslash \{0\}$ due to Claim 6.13 and Lemma 7.13.

Step 5: Smoothing of C_1: We apply Lemma 7.4 with $\tilde{\alpha}(x) = C_1(x)$, $\mu(x) = \frac{1}{2}C_1(x)$ and $\nu(x) = \frac{1}{4}C_1(x)$ and with respect to the open set $\mathbb{R}^n \backslash \{0\}$. Then, $C_2 : \mathbb{R}^n \to \mathbb{R}_{\geq 0}$ obtained through Lemma 7.4 is smooth for $x \neq 0$ and continuous at $x = 0$. Additionally C_2 satisfies

$$\tfrac{1}{2}\alpha_1(|x|) \leq \tfrac{1}{2}C_1(x) \leq C_2(x) \leq \tfrac{3}{2}C_1(x) \leq \tfrac{3}{2}\alpha_2(|x|) \tag{6.52}$$

which follows from (6.32) and (7.11), i.e., $|C_1(x) - C_2(x)| \leq \frac{1}{2}C_1(x)$. From (6.33), (7.12) and (6.52) additionally the bound

$$\min_{w \in F_L(x)} \langle \nabla C_2(x), w \rangle \geq \tilde{\alpha}(x) - \tfrac{1}{4}\nu(x) = \tfrac{1}{2}\tfrac{3}{2}C_1(x) \geq \tfrac{1}{2}C_2(x) \tag{6.53}$$

is obtained.

Finally, to get a function that is smooth on $\mathbb{R}^n$, we apply Lemma 7.14 to the function C_2. In particular, according to Lemma 7.14 there exists a smooth function $\rho \in \mathcal{K}_\infty$ with $\rho(s) \leq s\rho'(s)$ and $\rho' \in \mathcal{K}_\infty$ such that $C = (\rho \circ C_2)^2$ is smooth by construction. From (6.52) it follows that

$$\tilde{\alpha}_1(|x|) \leq C(x) \leq \tilde{\alpha}_2(|x|)$$

where

$$\tilde{\alpha}_1(s) = \rho \left(\tfrac{1}{2} \alpha_1(s) \right)^2, \qquad \tilde{\alpha}_2(s) = \rho \left(\tfrac{3}{2} \alpha_2(s) \right)^2.$$

Also, from (6.53) and the relation $\rho(s) \leq s\rho'(s)$, it follows that

$$
\begin{aligned}
\min_{w \in F_L(x)} \langle \nabla C(x), w \rangle &= \min_{w \in F(x)} \langle \nabla (\rho(C_2(x))^2, w \rangle \\
&= \min_{w \in F(x)} \langle 2\rho(C(x))\rho'(C_2(x))\nabla C_2(x), w \rangle \\
&\geq \left(2\rho(C_2(x))\rho'(C_2(x)) \right) \cdot \tfrac{1}{2} C_2(x) \\
&= \rho(C_2(x)) \cdot \left(\rho'(C_2(x))C_2(x) \right) \\
&\geq \rho(C_2(x))^2 = C(x).
\end{aligned}
$$

Thus, C is a smooth Chetaev function which completes the proof. $\square$

6.3 Proof of Theorem 4.11

In this section we show that weak $\mathcal{K}_\infty \mathcal{K}_\infty$-instability of the origin according to Definition 4.8 is equivalent to the existence of a control Chetaev function according to Definition 4.10. The sufficiency part of the theorem, i.e., the implication that a Lipschitz continuous control Chetaev function implies $\mathcal{K}_\infty \mathcal{K}_\infty$-instability, is already contained in the conference paper [11] for Lipschitz control Chetaev functions. Here we provide a proof that works for merely continuous control Chetaev functions, by adapting a technique from [22] from control Lyapunov to control Chetaev functions.

6.3.1 Control Chetaev Function Implies Weak Complete Instability

We start with a preparatory lemma and then turn to the proof of the result itself.

Lemma 6.14 *Consider the differential inclusion (2.1) satisfying Assumptions 2.1 and 2.2. Let C be a continuous control Chetaev function according to Definition 4.10. Then there are $\tilde{\alpha}_1, \tilde{\alpha}_2 \in \mathcal{K}_\infty$ and $\tilde{\rho} \in \mathcal{P}$ such that for all $\Delta > \delta > 0$ there is a Lipschitz continuous function $\tilde{C}$, which satisfies*

$$\tilde{\alpha}_1(|x|) \leq \tilde{C}(x) \leq \tilde{\alpha}_2(|x|) \tag{6.54}$$

$$\max_{w \in F(x)} D^+ \tilde{C}(x; w) \geq \tilde{\rho}(|x|) \tag{6.55}$$

for all $x \in \mathbb{R}^n$ with $\delta \leq |x| \leq \Delta$.

Proof By rescaling C if necessary (similar to the scaling in Remark 3.17) we can assume that the upper bound α_2 of C satisfies $\alpha_2(r) \leq r^2$. Moreover, we may assume that ρ, α_1 and α_2 in Definition 4.10 are Lipschitz (see [36, Lemma 1]).

We claim that $\tilde{C} = C_\beta$ for the sup-convolution

$$C_\beta(x) := \sup_{y \in \mathbb{R}^n} \left\{ C(y) - \frac{|x - y|^2}{2\beta^2} \right\}$$

satisfies the assertion for all $\beta > 0$ sufficiently small. For proving this claim, denote by $y_\beta(x)$ the maximizer of the expression defining C_β for $x \in \mathbb{R}^n$.

Note that this maximizer exists whenever $2\beta^2 < 1$, since from the definition of C_β it follows that $C_\beta(x) \geq C(x) \geq 0$, while the expression in braces becomes negative for $|y| > |x|/(1 - \sqrt{2}\beta)$. Indeed, the condition $|y| > |x|/(1 - \sqrt{2}\beta)$ implies $|y - x| \geq |y| - |x| > \sqrt{2}\beta|y|$ and thus $|y|^2 - |y - x|^2/(2\beta^2) < 0$. Thus, the negativity of the term in the braces follows from the assumption that $\alpha_2(r) \leq r^2$. Hence, the supremum is attained on a compact set and since all expressions are continuous it is indeed a maximum. In the remainder of the proof we restrict ourselves to $\beta > 0$ with $\sqrt{2}\beta < 1/2$, implying

$$|y_\beta(x)| \leq 2|x|. \tag{6.56}$$

Because of

$$
\begin{aligned}
C_\beta(x_1) - C_\beta(x_2) &= C(y_\beta(x_1)) - \frac{|x_1 - y_\beta(x_1)|^2}{2\beta^2} - \sup_{y \in \mathbb{R}^d} \left\{ C(y) - \frac{|x_2 - y|^2}{2\beta^2} \right\} \\
&\leq C(y_\beta(x_1)) - \frac{|x_1 - y_\beta(x_1)|^2}{2\beta^2} - C(y_\beta(x_1)) + \frac{|x_2 - y_\beta(x_1)|^2}{2\beta^2} \\
&= \frac{|x_2 - y_\beta(x_1)|^2 - |x_1 - y_\beta(x_1)|^2}{2\beta^2} \\
&= \frac{|x_1|^2 - |x_2|^2 + 2\langle x_2 - x_1, y_\beta(x_1) \rangle}{2\beta^2} \\
&\leq \frac{2\max\{|x_1|, |x_2|\}|x_1 - x_2| + 4|x_1||x_1 - x_2|}{2\beta^2},
\end{aligned}
$$

the function C_β is locally Lipschitz.

Defining

$$\zeta_\beta(x) := \frac{y_\beta(x) - x}{\beta^2},$$

by a slight adaptation of [22, Eq. (10) and Lemma III.2] (or by direct computations) one obtains

$$C_\beta(x + \tau v) \geq C_\beta(x) + \tau \langle \zeta_\beta(x), v \rangle - \frac{\tau^2 |v|^2}{2\beta^2} \tag{6.57}$$

and

$$C(y_\beta(x) + \tau v) \leq C(y_\beta(x)) + \tau \langle \zeta_\beta(x), v \rangle + \frac{\tau^2 |v|^2}{2\beta^2}.$$

The second inequality implies

$$\langle \zeta_\beta(x), v \rangle \geq \frac{C(y_\beta(s) + \tau v) - C(y_\beta(x))}{\tau} - \frac{\tau |v|^2}{2\beta^2}.$$

For any $v \in \mathbb{R}^n$ and all sequences $v_k \to v$ and $\tau_k \to 0$ we thus obtain

$$\langle \zeta_\beta(x), v \rangle = \limsup_{k \to \infty} \langle \zeta_\beta(x), v_k \rangle \geq \limsup_{k \to \infty} \frac{C(y_\beta(s) + \tau_k v_k) - C(y_\beta(x))}{\tau_k},$$

which implies

$$\langle \zeta_\beta(x), v \rangle \geq D^+ C(\zeta_\beta(x); v). \tag{6.58}$$

Now let $\omega \in \mathcal{K}_\infty$ be a modulus of continuity of C on $\{x \in \mathbb{R}^n \mid |x| \leq 2\Delta\}$, i.e., $|C(x_1) - C(x_2)| \leq \omega(|x_1 - x_2|)$ for all $x_1, x_2 \in \mathbb{R}^n$ with $|x_1| \leq 2\Delta$ and $|x_2| \leq 2\Delta$. Let $x \in \mathbb{R}^n$ be arbitrary with $\delta \leq |x| \leq \Delta$. The condition on x implies that $|y_\beta(x)| \leq 2\Delta$ according to (6.56). Then from the definition of C_β and y_β we obtain

$$\frac{|y_\beta(x) - x|^2}{2\beta^2} = C(y_\beta(x)) - C_\beta(x) \leq C(y_\beta(x)) - C(x) \leq \omega(|y_\beta(x) - x|).$$

Since (6.56) implies $|y_\beta(x)| \leq 2\Delta$ and $|y_\beta(x) - x| \leq 3|x| \leq 3\Delta$, we obtain

$$|y_\beta(x) - x| \leq \beta \sqrt{2\omega(3\Delta)}.$$

Together with the definition of ζ_β the last two inequalities imply

$$|\zeta_\beta(x)||y_\beta(x) - x| = 2\frac{|y_\beta(x) - x|^2}{2\beta^2} \leq 2\omega(|y_\beta(x) - x|) \leq 2\omega(\beta\sqrt{2\omega(3\Delta)}).$$

Moreover, since $C(x) \leq C_\beta(x) \leq C(y_\beta(x))$ we obtain

$$|C_\beta(x) - C(x)| \leq C(y_\beta(x)) - C(x) \leq \omega(|y_\beta(x) - x|) \leq \omega(\beta\sqrt{2\omega(3\Delta)}). \tag{6.59}$$

Observe that the right hand sides of these inequalities tend to 0 as $\beta \to 0$.

Now, pick $w \in F(y_\beta(x))$ with $D^+ C(y_\beta(x); v) \geq \rho(|y_\beta(x)|)$, which by (6.58) implies $\langle \zeta_\beta(x), w \rangle \geq \rho(|y_\beta(x)|)$. Since F is Lipschitz, we find $w' \in F(x)$ with $|w - w'| \leq L_F |y_\beta(x) - x|$, implying

$$\langle \zeta_\beta(x), w' \rangle \geq \langle \zeta_\beta(x), w \rangle - L_F |\zeta_\beta(x)||y_\beta(x) - x|$$
$$\geq \rho(|y_\beta(x)|) - 2L_F \omega(\beta\sqrt{2\omega(3\Delta)})$$
$$\geq \rho(|x|) - L_\rho \beta\sqrt{2\omega(3\Delta)} - 2L_F \omega(\beta\sqrt{2\omega(3\Delta)}),$$

where L_ρ denotes a Lipschitz constant for ρ. By choosing $\beta > 0$ so small that the sum of the last two terms is less than $\rho(\delta)/2$, for all $|x| \geq \delta$ we can conclude

$$\langle \zeta_\beta(x), w' \rangle \geq \rho(|x|)/2.$$

Setting $\tilde{\rho} = \rho/2$, using (6.57) we thus get

$$D^+ C_\beta(x; w') \geq \lim_{\tau \to 0} \frac{C_\beta(x + \tau w') - C_\beta(x)}{\tau}$$
$$\geq \langle \zeta_\beta(x), w' \rangle - \lim_{\tau \to 0} \frac{\tau |w'|}{2\beta^2} \geq \tilde{\rho}(|x|).$$

This shows (6.55) for $\tilde{C} = C_\beta$. The upper and lower bounds in (6.54) follow from (6.59) for all β sufficiently small with $\tilde{\alpha}_1 = \alpha_1/2$ and $\tilde{\alpha}_2 = 2\alpha_2$. $\qquad \square$

Proposition 6.15 *Consider the differential inclusion (2.1) satisfying Assumptions 2.1 and 2.2. Let C be a continuous control Chetaev function according to Definition 4.10. Then, (2.1) is weakly $\mathcal{K}_\infty \mathcal{K}_\infty$-unstable according to Definition 4.8.*

Proof By Lemma 6.14 we can assume that $C = \tilde{C}$ is Lipschitz for all x with $\delta < |x| < \Delta$. We keep writing C, ρ and α_i instead of $\tilde{C}$, $\tilde{\rho}$ and $\tilde{\alpha}_i$ in this proof in order to simplify the notation. Note that the comparison functions ρ and α_i provided by Lemma 6.14 do not depend on δ and Δ.

Using Theorem 2.3 in combination with Corollary 2.4, we can assume without loss of generality that all solutions are forward complete and, without loss of generality, satisfy $|\phi(t; x) - x| \leq t$ for all $t \geq 0$.

Fixing arbitrary $\Delta > \delta > 0$, we first prove that for each $x \in \mathbb{R}^n$ there is a solution $\psi \in S(x)$ such that

$$C(\psi(t; x)) \geq C(\psi(s; x)) + \frac{1}{2} \int_s^t \rho(|\psi(\tau; x)|) d\tau \qquad (6.60)$$

holds for all $t > s \geq 0$ with $\delta \leq |\psi(\tau; x)| \leq \Delta$ for all $\tau \in [s, t]$. Note that for $x = 0$ this property is trivially satisfied through $\psi(\cdot; 0) \equiv 0$ and thus we can assume $x \neq 0$ in the following, implying $|x| > \delta$ for all $\delta > 0$ sufficiently small. Since F is Lipschitz, it follows from Lemma 7.12 that for each $w \in F(x)$ there is a solution $\phi \in S(x)$ with

$$\frac{\phi(t; x) - x}{t} \to w.$$

Now, let $w \in F(x)$ be a vector for which $D^+ C(x; w) \geq \rho(|x|)$ holds, let $t_n \searrow 0$ be a sequence for which the limit superior in the definition of the Dini derivative D^+ is obtained and set $v_k = \frac{1}{t_k}(\phi(t_k; x) - x)$.

Since C is Lipschitz, the limit superior in the definition of D^+ does not depend on the choice of the sequence v_k. We thus obtain

$$\rho(|x|) \leq D^+ C(x; w) = \lim_{k \to \infty} \frac{C(x + t_k v_k) - C(x)}{t_k} = \lim_{k \to \infty} \frac{C(\phi(t_k; x)) - C(x)}{t_k}.$$

Hence, for k sufficiently large we obtain

$$C(\phi(t_k; x)) \geq C(x) + \frac{3}{4} t_k \rho(|x|) = C(x) + \frac{3}{4} \int_0^{t_k} \rho(|x|) d\tau.$$

Since ρ and $t \mapsto \phi(t; x)$ are continuous, again for k sufficiently large we obtain $\rho(|\phi(t; x)|) \geq \frac{2}{3} \rho(|x|)$ for all $t \in [0, t_k]$, and we get

$$C(\phi(t_k; x)) \geq C(x) + \frac{3}{4} \int_0^{t_k} \frac{2}{3} \rho(|\phi(\tau; x)|) d\tau = C(x) + \frac{1}{2} \int_0^{t_k} \rho(|\phi(\tau; x)|) d\tau$$

for all sufficiently large k. In other words, for each $x \in \mathbb{R}^n$ and each $\gamma > 0$ there is $t_\gamma \in (0, \gamma]$ and $\phi_\gamma \in S(x)$ with

$$C(\phi(t_\gamma; x)) \geq C(x) + \frac{1}{2} \int_0^{t_\gamma} \rho(|\phi(\tau; x)|) d\tau. \tag{6.61}$$

Now fix $\gamma > 0$ and proceed inductively as follows:

Set $i := 0$, $\tau_0 := 0$ and $x_0 := x$. Then, for $i = 1, 2, 3, \ldots$ let τ_i be equal to $\tau_{i-1} + t_\gamma$ with t_γ being the maximal $t_\gamma \leq \gamma$ satisfying (6.61) for $x = x_{i-1}$, and set $x_i := \phi_\gamma(\tau_i - t_\gamma; x_{i-1})$.

This yields an increasing sequence of times τ_i with $\tau_{i+1} - \tau_i < \gamma$ and by concatenating the solutions $\phi_\gamma(\tau_i - t_\gamma; x_{i-1})$ on the subintervals we obtain a solution $\psi_\gamma(t; x)$ that satisfies (6.60) for all $t = \tau_k$ and $s = \tau_m, m < k$. We now prove $\tau_k \to \infty$ by contradiction. Assume that this is not the case and let $\tau_{\max} = \lim_{k \to \infty} \tau_k$. Then by continuity of all involved expressions in t, the solution $\psi(t; x)$ satisfies (6.60) also for $t = \tau_{\max}$. Then, using t_γ from (6.61) with $\gamma/2$ in place of γ, the solution can be extended to satisfy (6.60) also at time $t = t_{\max} + t_\gamma$. This, however, implies that for all sufficiently large k the τ_k could have been chosen to satisfy $\tau_k \geq t_{\max} + t_\gamma$, contradicting their maximality.

Now, fix an arbitrary $T > 0$ and let L denote the Lipschitz constant of C on $\overline{B}_T(x)$. Then for the solution ψ_γ just constructed and any $s \in [\tau_m, \tau_{m+1}], t \in [\tau_k, \tau_{k+1}]$ with $s < t$ and $\tau_{k+1} \leq T$ we can estimate

$$C(\psi_\gamma(t;x)) \geq C(\psi_\gamma(\tau_{k+1};x)) - L\gamma$$

$$\geq C(\psi_\gamma(\tau_m;x)) + \frac{1}{2}\int_{\tau_m}^{t_{k+1}} \rho(|\psi_\gamma(\tau;x)|)d\tau - L\gamma$$

$$\geq C(\psi_\gamma(s;x)) + \frac{1}{2}\int_s^t \rho(|\psi_\gamma(\tau;x)|)d\tau - 2L\gamma.$$

Now we consider a sequence $\gamma_k \to 0$ and the corresponding sequence ψ_{γ_k}. As in the proof of "(a) $\Rightarrow$ (b)" of Lemma 6.1, by Lemma 7.7 we can assume that this sequence converges to a limit solution $\psi \in S(x)$, uniformly on any compact time interval. This limit solution satisfies the above inequality with $\gamma = 0$ for all $s, t \in [0, T]$ with $\delta \leq |\psi(\tau;x)| \leq \Delta$ for all $\tau \in [s, t]$.

Now $t \mapsto C(\psi(t;x))$ is a Lipschitz and hence absolutely continuous function. It is thus differentiable for almost any $t \geq 0$ and due to (6.60) it satisfies

$$\frac{d}{dt}C(\psi(t;x)) \geq \frac{1}{2}\rho(|\psi(t,x)|) \geq \frac{1}{2}\eta(C(\psi(t,x))) \qquad (6.62)$$

for $\eta = \rho \circ \alpha_2^{-1}$ at all points of differentiability $t \in [0, T]$ with $\delta \leq |\psi(t,x)| \leq \Delta$. Hence, Lemma 7.1 applied to (6.62) provides a function $\kappa \in \mathcal{K}_\infty\mathcal{K}_\infty$ such that $C(\psi(t;x)) \geq \kappa(C(x), t)$ and the bounds (4.7) lead to

$$|\psi(t;x)| \geq \alpha_2^{-1} \circ \kappa(\alpha_1(|x|), t) = \tilde{\kappa}(|x|, t) \qquad (6.63)$$

with $\tilde{\kappa} = \alpha_2^{-1} \circ \kappa(\alpha_1(\cdot), \cdot) \in \mathcal{K}_\infty\mathcal{K}_\infty$. Choosing $\delta > 0$ so small that $\alpha_2(\delta) < \alpha_1(|x|)$, since C is increasing along the ψ, we can exclude the case that $|\psi(\tau, x)| < \delta$, hence the inequality holds as long as $t \leq T$ and $|\psi(t, x)| \leq \Delta$, i.e., at least for all $t \in [0, \min\{\Delta - |x|, T\}]$.

Now, finally, we consider a sequence $\Delta_k \to \infty$ and the corresponding solutions ψ_k with $T = T_k = \Delta_k - |x|$, i.e., satisfying (6.63) for all $t \in [0, \Delta_k - |x|]$. Again we find a subsequence that converges to a solution ϕ uniformly on compact time intervals, implying that ϕ satisfies (6.63) for all $t \geq 0$. This finishes the proof. $\qquad \square$

6.3.2 Weak Complete Instability Implies the Existence of a Control Chetaev Function

We continue with the converse direction of Theorem 4.11. The statement is repeated for convenience here.

Proposition 6.16 *Consider the differential inclusion (2.1) satisfying Assumptions 2.1 and 2.2. Let (2.1) be weakly $\mathcal{K}_\infty\mathcal{K}_\infty$-unstable according to Definition 4.8. Then there exists a continuous control Chetaev function according to Definition 4.10.*

Proof We start with some preliminary observations. First, without loss of generality we assume that the differential inclusion (2.1) satisfies

$$F(x) \subset \bar{B}_1(0) \qquad \forall x \in \mathbb{R}^n. \tag{6.64}$$

If the statement of Proposition 6.16 is satisfied under this additional condition, then the general results follows from Theorem 2.3 and Lemma 4.21.

The Assumption (6.64) implies that for each $x \in \mathbb{R}^n$ all solutions $\phi \in \mathcal{S}(x)$ satisfy the bounds

$$|x| - t \le |\phi(t; x)| \le |x| + t, \qquad \forall t \in \mathbb{R}_{\ge 0}. \tag{6.65}$$

Furthermore, for all $x \in \mathbb{R}^n$ weak complete instability guarantees the existence of a solution $\phi \in \mathcal{S}(x)$ such that

$$|\phi(t; x)| \ge \kappa(|x|, t), \qquad \forall t \in \mathbb{R}_{\ge 0}, \tag{6.66}$$

and for $\kappa \in \mathcal{K}_\infty \mathcal{K}_\infty$. Combined with Lemma 7.2, for $\lambda = 1$, there exist $\alpha_1, \alpha_2 \in \mathcal{K}_\infty$ such that for all $x \in \mathbb{R}^n$ there exists $\phi \in \mathcal{S}(x)$ satisfying the lower bound

$$|\phi(t; x)| \ge \alpha_2(\alpha_1(|x|)e^t), \qquad \forall t \in \mathbb{R}_{\ge 0}. \tag{6.67}$$

To define a control Chetaev function we consider the cost functional $J(\cdot, \cdot) : \mathbb{R}^n \times \mathcal{S}(x) \to \mathbb{R}_{\ge 0}$,

$$J(x, \phi) = \begin{cases} \frac{1}{\int_0^\infty \frac{1}{g(\phi(t;x))} dt}, & \text{if } \int_0^\infty g(\phi(t; x))^{-1} dt \text{ exists,} \\ 0, & \text{otherwise,} \end{cases} \tag{6.68}$$

defined through a continuous function $g : \mathbb{R}^n \to \mathbb{R}$ satisfying the bounds

$$\gamma_1(|x|) \le g(x) \le \gamma_2(|x|) \tag{6.69}$$

for $\mathcal{K}_\infty$-functions $\gamma_1, \gamma_2 \in \mathcal{K}_\infty$. Here we select g, γ_1 and γ_2 such that the (nonrestrictive) conditions

$$\gamma_1(r) \le \gamma_2(r) \le r \qquad \forall r \in [0, 2], \tag{6.70}$$

$$\alpha_2^{-1}(r) \le \gamma_1(r) \le \gamma_2(r) \qquad \forall r \in [3, \infty), \tag{6.71}$$

are satisfied. Note that the functional J is well-defined. Indeed, since g is nonnegative and $t \mapsto g(\phi(t; x))^{-1}$ is continuous, except for the times t where $g(\phi(t; x)) = 0$, existence of $\int_0^\infty g(\phi(t; x))^{-1} dt$ is equivalent to the conditions that the set of $t \ge 0$ with $g(\phi(t; x)) = 0$ has measure 0 and $\lim_{T \to \infty} \int_0^T g(\phi(t; x))^{-1} dt < \infty$. In both cases, $J(x, u) = 0$ is consistent with the interpretations "$\frac{1}{0} = \infty$" and "$\frac{1}{\infty} = 0$".

Based on J we define the optimal value function

$$C(x) = \sup_{\phi \in S(x)} J(x, \phi) \qquad (6.72)$$

which satisfies $C : \mathbb{R}^n \to \mathbb{R}_{\geq 0}$. Equivalently, for $C(x) \neq 0$, it holds that

$$\frac{1}{C(x)} = \inf_{\phi \in S(x)} \frac{1}{J(x, \phi)}. \qquad (6.73)$$

The selection of J and the definition of C are motivated through the following claim.

Claim 6.17 *Assume the function C in (6.72) is continuous and satisfies the bounds*

$$\tilde{\alpha}_1(|x|) \leq C(x) \leq \tilde{\alpha}_2(|x|) \qquad \forall \, x \in \mathbb{R}^n. \qquad (6.74)$$

Then there exists $\tilde{\rho} \in \mathcal{P}$ such that the increase condition

$$\max_{w \in F(x)} D^+ C(x, w) \geq \tilde{\rho}(|x|) \qquad (6.75)$$

is satisfied for all $x \in \mathbb{R}^n$.

Proof of Claim 6.17 Since $C(0) = 0$ and $C(x) \geq 0$ for all $x \in \mathbb{R}^n$, the definition of the Dini derivative in Eq. (2.19a) implies that $D^+ C(0, w) \geq 0$ for all $w \in \mathbb{R}^n$. Thus, (6.75) is trivially satisfied for $x = 0$ and for all positive definite functions $\tilde{\rho} \in \mathcal{P}$.
   Let $x \neq 0$. Then, according to (6.65),

$$0 < |\phi(t; x)| < \infty \qquad \forall t \in [0, \infty), \, \forall \phi \in S(x). \qquad (6.76)$$

Since $x \neq 0$, we can use the representation (6.73) and for $T \geq 0$ it holds that

$$\begin{aligned}
\frac{1}{C(x)} &= \inf_{\phi \in S(x)} \int_0^\infty \frac{1}{g(\phi(t; x))} dt \\
&= \inf_{\phi \in S(x)} \left(\int_0^T \frac{1}{g(\phi(t; x))} dt + \int_T^\infty \frac{1}{g(\phi(t; x))} dt \right) \\
&\geq \inf_{\phi \in S(x)} \left(\int_0^T \frac{1}{g(\phi(t; x))} dt + \frac{1}{C(\phi(T; x))} \right)
\end{aligned}$$

We select a sequence of solutions $(\psi_k)_{k \in \mathbb{N}} \subset S(x)$ such that

$$\frac{1}{C(x)} \geq \left(1 - \tfrac{1}{k}\right) \int_0^{T(x)} \frac{1}{g(\psi_k(t; x))} dt + \frac{1}{C(\psi_k(T(x); x))}. \qquad (6.77)$$

Since all solutions are bounded according to (6.76) we can apply Lemma 7.7 for $\mathcal{A} = \{x\}$ and $T = T(x)$ and thus $(\psi_k)_{k \in \mathbb{N}}$ converges uniformly on $[0, T(x)]$ to $\psi \in$

$S(x)$ and continuity of g and C imply that

$$\frac{1}{C(x)} \geq \int_0^{T(x)} \frac{1}{g(\psi(t;x))} dt + \frac{1}{C(\psi(T(x);x))}. \tag{6.78}$$

Thus, by the dynamic programming principle $\psi \in S(x)$ is an optimal solution on $[0, T(x)]$ and using the dynamic programming principle once more implies that

$$\frac{1}{C(x)} = \int_0^T \frac{1}{g(\psi(t;x))} dt + \frac{1}{C(\psi(T;x))}. \tag{6.79}$$

for all $T \in [0, T(x)]$. Rearranging terms and dividing by $T > 0$ we obtain

$$\begin{aligned}
0 &= \frac{1}{T} \int_0^T \frac{1}{g(\psi(t;x))} dt + \frac{1}{T} \left(\frac{1}{C(\psi(T;x))} - \frac{1}{C(x)} \right) \\
&= \frac{1}{T} \int_0^T \frac{1}{g(\psi(t;x))} dt + \frac{1}{T} \frac{C(x) - C(\psi(T;x))}{C(\psi(T;x))C(x)}.
\end{aligned} \tag{6.80}$$

Since $\dot{\psi}(t;x) \in F(\psi(t;x))$ for almost all $t \geq 0$ (and F is locally Lipschitz continuous and convex valued and ψ is absolutely continuous), there exists $t_1 \in (0, T(x)]$ such that $\psi(t;x)$ can be written as

$$\psi(t;x) = x + t(w + r(t)), \qquad \forall t \in [0, t_1)$$

for $w \in F(x)$ and a continuous function $r : [0, t_1) \to \mathbb{R}^n$ satisfying $\lim_{t \searrow 0} r(t) = 0$. Moreover, for $x \neq 0$, the function $\tilde{C}(x) = C(x)^{-1}$ satisfies the following property with respect to the Dini derivative:

$$\begin{aligned}
D_+ \tilde{C}(x, w) &= \liminf_{v \to w; \, t \searrow 0} \frac{1}{t} \left(\frac{1}{C(x+tv)} - \frac{1}{C(x)} \right) = \liminf_{v \to w; \, t \searrow 0} \frac{1}{t} \left(\frac{C(x) - C(x+tv)}{C(x+tv)C(x)} \right) \\
&= \frac{1}{C(x)^2} \liminf_{v \to w; \, t \searrow 0} -\frac{1}{t} (C(x+tv) - C(x)) \\
&= -\frac{1}{C(x)^2} D^+ C(x, w).
\end{aligned}$$

Combining these properties and taking the limit $T \to 0$ in (6.80) it holds that

$$\begin{aligned}
0 &= \lim_{T \to 0} \frac{1}{T} \int_0^T \frac{1}{g(\psi(t;x))} dt + \liminf_{v \to w; \, T \searrow 0} \frac{1}{T} \left(\frac{C(x) - C(x + T(v + r(T)))}{C(x + T(v + r(T)))C(x)} \right) \\
&= \frac{1}{g(\psi(0;x))} - \frac{1}{C(x)^2} D^+ C(x; w)
\end{aligned}$$

and thus

$$\sup_{w \in F(x)} D^+ C(x; w) \geq \frac{C(x)^2}{g(x)}.$$

To complete the proof, we define $\rho : \mathbb{R}_{\geq 0} \to \mathbb{R}_{\geq 0} \cup \{\infty\}$,

$$\rho(r) = \inf_{|x|=r} C(x)^2 g(x)^{-1} \quad \forall r \in \mathbb{R}_{>0} \quad \text{and}$$

$$\rho(0) = \lim_{r \to 0} \inf_{|x|=r} C(x)^2 g(x)^{-1}.$$

The function satisfies $\rho(r) > 0$ for $r > 0$ and $\rho(r) \geq 0$ for $r = 0$. Moreover, since C and g are continuous, ρ is continuous for all r such that $\rho(r) < \infty$ holds. Thus there exists a continuous function $\tilde{\rho} \in \mathcal{P}$ such that $\tilde{\rho}(r) \leq \rho(r)$ and C satisfies $\sup_{w \in F(x)} D^+ C(x; w) \geq \tilde{\rho}(|x|)$, i.e., the increase condition (6.75) is satisfied for all $x \in \mathbb{R}^n$. □

If the assumptions of Claim 6.17 are satisfied, then C is a control Chetaev function according to Definition 4.10. To complete the proof of the theorem, we need to show that C satisfies the bounds (6.74) and is continuous.

Claim 6.18 *The function C defined in (6.72) satisfies (6.74).*

Proof of Claim 6.18 For $x = 0$ the bounds are again trivially satisfied. For $x \neq 0$ we can rewrite the conditions (6.69), (6.70) and (6.69) to

$$\frac{1}{\gamma_2(|x|)} \leq \frac{1}{g(x)} \leq \frac{1}{\gamma_1(|x|)} \quad \forall r \in (0, \infty), \tag{6.81}$$

$$\frac{1}{r} \leq \frac{1}{\gamma_2(r)} \leq \frac{1}{\gamma_1(r)} \quad \forall r \in (0, 2], \tag{6.82}$$

$$\frac{1}{\gamma_2(r)} \leq \frac{1}{\gamma_1(r)} \leq \frac{1}{\alpha_2^{-1}(r)} \quad \forall r \in [3, \infty), \tag{6.83}$$

Moreover, all solutions $\phi \in S(x)$ satisfy

$$\frac{1}{|x| + t} \leq \frac{1}{|\phi(t; x)|}, \quad \forall t \in \mathbb{R}_{\geq 0}, \tag{6.84}$$

and there exist $\phi \in S(x)$ such that

$$\frac{1}{\alpha_2(\alpha_1(|x|)e^t)} \geq \frac{1}{|\phi(t; x)|}, \quad \forall t \in \mathbb{R}_{\geq 0}. \tag{6.85}$$

Using (6.84), for $x \neq 0$ and $\phi \in S(x)$ arbitrary, it holds that

$$\frac{1}{J(x,\phi)} = \int_0^\infty \frac{1}{g(\phi(t;x))}dt \geq \int_0^1 \frac{1}{g(\phi(t;x))}dt \geq \int_0^1 \frac{1}{\gamma_2(|\phi(t;x)|)}dt$$

$$\geq \int_0^1 \frac{1}{\gamma_2(|x|+t)}dt \geq \int_0^1 \frac{1}{\gamma_2(|x|+1)}dt = \frac{1}{\gamma_2(|x|+1)}. \tag{6.86}$$

Moreover, for $|x| \leq 1$, inequality (6.65) implies that $|\phi(t;x)| \leq 2$ for all $t \leq 1$ and for all $\phi \in S(x)$. Thus, for $x \in \bar{B}_1(0)\backslash\{0\}$, inequality (6.82) is applicable and the estimate (6.86) can be refined to

$$\frac{1}{J(x,\phi)} \geq \int_0^1 \frac{1}{\gamma_2(|x|+t)}dt \geq \int_0^1 \frac{1}{|x|+t}dt = \ln\left(\frac{|x|+1}{|x|}\right).$$

Since these estimates are satisfied for all $\phi \in S(x)$, it holds that

$$\frac{1}{C(x)} \geq \frac{1}{\gamma_2(|x|+1)} \quad \forall x \in \mathbb{R}^n\backslash\{0\}, \tag{6.87}$$

$$\frac{1}{C(x)} \geq \ln\left(\frac{|x|+1}{|x|}\right) \quad \forall x \in \bar{B}_1(0)\backslash\{0\}, \tag{6.88}$$

or equivalently

$$C(x) \leq \gamma_2(|x|+1) \quad \forall x \in \mathbb{R}^n\backslash\{0\}, \tag{6.89}$$

$$C(x) \leq \ln\left(\frac{|x|+1}{|x|}\right)^{-1} \quad \forall x \in \bar{B}_1(0)\backslash\{0\}. \tag{6.90}$$

The function $\chi : \mathbb{R}_{>0} \to \mathbb{R}_{>0}$, $\chi(r) = \ln(r^{-1}+1)$, is continuous, strictly monotonically decreasing and satisfies

$$\chi(r) \xrightarrow{r\to 0} \infty \quad \text{and} \quad \chi(r) \xrightarrow{r\to\infty} 0.$$

Thus, the function

$$\tilde{\chi}(r) = \begin{cases} 0, & r = 0, \\ \frac{1}{\chi(r)}, & r > 0, \end{cases}$$

is a $\mathcal{K}_\infty$-function. Moreover, since $\gamma_2(\cdot + 1) - \gamma_2(1) \in \mathcal{K}_\infty$ and $C(0) = 0$ by definition, from (6.89) and (6.90) the existence of $\tilde{\alpha}_2 \in \mathcal{K}_\infty$ so that $C(x) \leq \tilde{\alpha}_2(|x|)$ for all $x \in \mathbb{R}^n$, follows. This establishes the upper bound in (6.74).

For $x \neq 0$ we define

$$\tau(|x|) = \min\left\{0, \ln\left(\frac{\alpha_2^{-1}(3)}{\alpha_1(|x|)}\right)\right\}.$$

Let $\phi \in S(x)$ be such that (6.67) is satisfied and let $t \geq \tau(|x|)$. Then it holds that

$$|\phi(t; x)| \geq \alpha_2(\alpha_1(|x|)e^t) \geq \alpha_2\left(\alpha_1(|x|)e^{\ln\left(\frac{\alpha_2^{-1}(3)}{\alpha_1(|x|)}\right)}\right) \geq \alpha_2\left(\alpha_1(|x|)\left(\frac{\alpha_2^{-1}(3)}{\alpha_1(|x|)}\right)\right) = 3.$$

In this case the solution can be written as

$$\frac{1}{J(x, \phi)} = \int_0^{\tau(|x|)} \frac{1}{g(\phi(t; \phi))}dt + \int_{\tau(|x|)}^\infty \frac{1}{g(\phi(t; \phi))}dt$$

and the first integral satisfies

$$\int_0^{\tau(|x|)} \frac{1}{g(\phi(t; \phi))}dt \leq \int_0^{\tau(|x|)} \frac{1}{\gamma_1(|\phi(t; \phi)|)}dt \leq \int_0^{\tau(|x|)} \frac{1}{\gamma_1(\alpha_2(\alpha_1(|x|)e^t))}dt$$

$$\leq \int_0^{\tau(|x|)} \frac{1}{\gamma_1(\alpha_2(\alpha_1(|x|)))}dt = \frac{\tau(|x|)}{\gamma_1(\alpha_2(\alpha_1(|x|)))}.$$

Here we have used the bounds in (6.81) and in (6.85). For the second integral, since $\gamma_1(r) \geq \alpha_2^{-1}(r)$ for $r \geq 3$, we obtain

$$\int_{\tau(|x|)}^\infty \frac{1}{g(\phi(t; x))}dt \leq \int_{\tau(|x|)}^\infty \frac{1}{\alpha_2^{-1}(|\phi(t; x)|)}dt \leq \int_{\tau(|x|)}^\infty \frac{1}{\alpha_2^{-1}(\alpha_2(\alpha_1(|x|)e^t)}dt$$

$$\leq \int_{\tau(|x|)}^\infty \frac{1}{\alpha_1(|x|)e^t}dt \leq \int_0^\infty \frac{1}{\alpha_1(|x|)e^t}dt = \frac{1}{\alpha_1(|x|)}.$$

Here we have used the bounds (6.83) and (6.85). Combining both estimates it holds that

$$\frac{1}{C(x)} \leq \frac{\tau(|x|)}{\gamma_1(\alpha_2(\alpha_1(|x|)))} + \frac{1}{\alpha_1(|x|)}$$

or equivalently

$$C(x) \geq \left(\frac{\tau(|x|)}{\gamma_1(\alpha_2(\alpha_1(|x|)))} + \frac{1}{\alpha_1(|x|)}\right)^{-1}.$$

The function $\tau(\cdot) : \mathbb{R}_{>0} \to \mathbb{R}_{\geq 0}$ is continuous and monotonically decreasing and the functions $\gamma_1(\alpha_2(\alpha_1(\cdot)))^{-1}, \alpha_1(\cdot)^{-1} : \mathbb{R}_{>0} \to \mathbb{R}_{>0}$ are strictly monotonically decreasing and satisfy

$$\gamma_1(\alpha_2(\alpha_1(r)))^{-1} \xrightarrow{r\to 0} \infty \quad \text{and} \quad \gamma_1(\alpha_2(\alpha_1(r)))^{-1} \xrightarrow{r\to\infty} 0,$$

$$\alpha_1(r)^{-1} \xrightarrow{r\to 0} \infty \quad \text{and} \quad \alpha_1(r)^{-1} \xrightarrow{r\to\infty} 0.$$

Thus $\tilde{\alpha}_1$ defined as

$$\tilde{\alpha}_1(r) = \begin{cases} \left(\frac{\tau(r)}{\gamma_1(\alpha_2(\alpha_1(r)))} + \frac{1}{\alpha_1(r)} \right)^{-1}, & r \neq 0, \\ 0, & r = 0, \end{cases}$$

is a $\mathcal{K}_\infty$-function and the lower bound in (6.74) follows.   □

Before we establish continuity of C, we show that the following auxiliary result is true. To this end, for $K > 0$, we define a subset of the solutions $S(x)$ as

$$S_K(x) = \{ \phi \in S(x) | \ J(x, \phi)^{-1} \leq K \}.$$

While $S_K(x)$ might be empty, the upper bound established in Claim 6.18 guarantees that $S_K(x) \neq \emptyset$ for $K \geq \tilde{\alpha}_2(|x|)$.

Claim 6.19 *Let $x \neq 0$ be arbitrary. For any $K > 0$ and any $R > 0$, there is a $T > 0$ such that for all $\phi \in S_K(x)$, there exists a $t_R(x) \leq T$, so that*

$$|\phi(t_R(x), x)| \geq R + 1. \tag{6.91}$$

Moreover, for all $\phi \in S_K(x)$, there exists $\eta > 0$ such that

$$|\phi(t; x)| \geq \eta, \qquad \forall t \geq 0. \tag{6.92}$$

Proof of Claim 6.19 Let $x \neq 0$ be arbitrary. Using inequality (6.81) the function g satisfies

$$\frac{1}{g(x)} > c \qquad \forall \ |x| < R + 1 \tag{6.93}$$

where c is defined as

$$c = \frac{1}{2} \inf_{0 < |x| < R+1} \frac{1}{\gamma_2(|x|)} = \frac{1}{2} \frac{1}{\sup_{|x| \leq R+1} \gamma_2(|x|)}.$$

Now, assume that (6.91) is not satisfied and in particular assume that $T \in \mathbb{R}_{>0}$ does not exist for a solution $\phi \in S_K(x)$. We can select $T = (K + 1)/c$ and by assumption $|\phi(t; x)| \leq R + 1$ is satisfied for all $t \in [0, T]$. Then (6.93) implies that $J(x, \phi)^{-1} \geq cT = K + 1$ and thus a contradiction to $\phi \in S_K(x)$. Hence, (6.91) is true.

To see that (6.92) is true, assume to the contrary that $|\phi(t; x)|$ becomes arbitrarily small. For all $s \in \mathbb{R}_{\geq 0}$ it holds that

$$\frac{1}{J(x, \phi)} = \int_0^s \frac{1}{g(|\phi(t; x)|)} dt + \int_s^\infty \frac{1}{g(|\phi(t; x)|)} dt$$

$$\geq \int_s^\infty \frac{1}{g(|\phi(t; x)|)} dt \geq \frac{1}{C(\phi(s; x))}$$

and thus

$$\frac{1}{J(x,\phi)} \geq \max_{s\geq 0} \frac{1}{C(\phi(s;x))}.$$

If $|\phi(t;x)|$ becomes arbitrarily small then there exists $s \in \mathbb{R}_{\geq 0}$ so that $|\phi(s,x)| \in B_1(0)$ and

$$\ln\left(\frac{|\phi(s,x)|+1}{|\phi(s,x)|}\right) > K.$$

Thus, (6.88) implies that

$$\frac{1}{J(x,\phi)} \geq \frac{1}{C(\phi(s,x))} \geq \ln\left(\frac{|\phi(s,x)|+1}{|\phi(s,x)|}\right) > K$$

which contradicts the assumption $\phi \in S_K(x)$. $\square$

With (6.91) and (6.92) we are able to establish continuity of $C(x)$.

Claim 6.20 *The function C defined in (6.72) is continuous.*

Proof of Claim 6.20 The proof of this claim follows ideas from [17]. Since the function C satisfies $0 \leq C(x) \leq \tilde{\alpha}_2(|x|)$ and $\tilde{\alpha}_2 \in \mathcal{K}_\infty$, continuity of C at $x = 0$ follows. To show continuity of C for $x \neq 0$, we focus on the function $C(x)^{-1}$. If $C(x)^{-1}$ is continuous for $x \neq 0$, then $C(x)$ is continuous for $x \neq 0$.

In order to prove continuity, we now show that for each $x \neq 0$ and each $\varepsilon > 0$, there is $\delta > 0$ such that

$$\left|\frac{1}{C(x_1)} - \frac{1}{C(x_2)}\right| \leq \varepsilon \qquad \forall x_1, x_2 \in B_\delta(x). \tag{6.94}$$

Fix $x \neq 0$ and $\varepsilon > 0$. We consider the ball $B_{\varepsilon_1}(x)$ with $\varepsilon_1 \leq \min\{|x|/2, 1\}$, set $R = \tilde{\alpha}_1^{-1}(\frac{\varepsilon}{2})$, implying $\frac{1}{C(x)} \leq \frac{\varepsilon}{2}$ for all $|x| \geq R$ (see (6.74) established in Claim 6.18) and define

$$K = \sup_{y \in B_{\varepsilon_1}(x)} \frac{1}{C(y)} + 1 = \frac{1}{\inf_{y \in B_{\varepsilon_1}(x)} C(y)} + 1.$$

Then $S_K(x) \neq \emptyset$ for all $B_{\varepsilon_1}(x)$ and we can find $T > 0$ and $\eta > 0$ which satisfy the properties of Claim 6.19 for all $x \in B_{\varepsilon_1}(x)$. On the compact set

$$Q = \overline{B}_{R+2}(0)\backslash B_{\frac{\eta}{2}}(0) = \left\{x \in \mathbb{R}^n | \tfrac{\eta}{2} \leq |x| \leq R+2\right\}$$

the function $1/g$ is uniformly continuous (since g is continuous and positive outside 0). Due to this fact, for each $\varepsilon_g > 0$, there exists $\delta_g > 0$ such that

$$\left| \frac{1}{g(x_1)} - \frac{1}{g(x_2)} \right| \leq \varepsilon_g, \qquad \forall \, x_1, x_2 \in Q \text{ with } |x_1 - x_2| \leq \delta_g. \tag{6.95}$$

We set $\varepsilon_g = \varepsilon/(2T)$ and we define the set

$$\mathcal{A} = \{z \in \mathbb{R}^n | z = \phi(t, y), t \in [0, T], y \in \overline{B}_{\varepsilon_1}(x), \phi \in \mathcal{S}(x)\} \backslash B_{\frac{\eta}{2}}(0).$$

Since $|\phi(t; x)| \leq |x| + t$ (see (6.65)), the set $\mathcal{A}$ is compact and excludes a neighborhood around the origin. According to Corollary 7.11 for any $\varepsilon_\phi > 0$ we can find $\delta_\phi > 0$ such that for each $\phi \in \mathcal{S}(x_1)$ we can find $\psi \in \mathcal{S}(x_2)$ such that

$$|\phi(t; x_1) - \psi(t; x_2)| \leq \varepsilon_\phi \quad \begin{array}{l} \forall \, x_1, x_2 \in B_{\varepsilon_1}(x) \text{ with } |x_1 - x_2| \leq \delta_\phi, \\ \forall \, t \in [0, T_{x_1}], \end{array} \tag{6.96}$$

where $T_{x_1} \in [0, T]$ is such that $\phi(t; x_1) \in \mathcal{A}$ for all $t \in [0, T_{x_1}]$.

We set $\varepsilon_\phi = \min\{1, \eta/2, \delta_g\}$. Through this selection, if $\phi \in \mathcal{S}_K(x_1)$ satisfies the properties of Claim 6.19, then $\phi(t; x) \in \mathcal{A}$ for all $t \in [0, T]$ and $T_{x_1} = T$ can be guaranteed. Indeed, $\phi(t; x_1) \in \mathcal{S}_K(x_1)$ implies that $|\phi(t; x_1)| \geq \eta$ for all $t \in [0, T]$, i.e., $\phi(t; x_1) \in \mathcal{A}$ for all $t \in [0, T]$. Moreover, $\varepsilon \leq \frac{1}{2}\eta$ implies that all $\psi \in \mathcal{S}(x_2)$ satisfying (6.96) additionally satisfy $|\psi(t; x_2)| \geq \frac{1}{2}\eta$.

We pick $x_1, x_2 \in B_{\delta_\phi}(x)$, fix $\varepsilon_2 \in (0, 1]$, and choose $\bar{\phi} \in \mathcal{S}(x_1)$ such that

$$\frac{1}{J(x_1, \bar{\phi})} \leq \frac{1}{C(x_1)} + \varepsilon_2.$$

By construction, there is $t_R(x_1) \leq T$ satisfying the properties of Claim 6.19 and we select $\bar{\psi} \in \mathcal{S}(x_2)$ such that (6.96) is satisfied. Then we have:

- $|\bar{\psi}(t; x_2)| > \eta/2$ for all $t \in [0, t_R(x_1)]$ since $t_R(x_1) \leq T$.
- $|\bar{\psi}(t_R(x_1), x_2)| \geq R$, since $|\bar{\phi}(t_R(x_1), x_1)| \geq R + 1$ (according to Claim 6.19) and $\varepsilon_\phi \leq 1$; and
- according to (6.95) and the definition of $\varepsilon_g = \varepsilon/(2T)$ it holds that

$$\left| \int_0^{t_R(x_1)} \frac{1}{g(\bar{\psi}(t; x_2))} dt - \int_0^{t_R(x_1)} \frac{1}{g(\bar{\phi}(t; x_1))} dt \right| \leq \frac{\varepsilon}{2}.$$

As a next step, we define $\bar{\varphi} \in \mathcal{S}(\bar{\psi}(t_R(x_1), x_2))$ as an ε_2-optimal solution, i.e.,

$$\frac{1}{C(\bar{\psi}(t_R(x_1), x_2))} \geq \frac{1}{J(\bar{\psi}(t_R(x_1), x_2), \bar{\varphi})} - \varepsilon_2.$$

Since we have not used the properties of $\bar{\psi}(t; x_2)$ for $t > t_R(x_1)$, we can assume without loss of generality that

$$\bar{\psi}(t + t_R(x_1), x_2) = \bar{\varphi}(t, \bar{\psi}(t_R(x_1), x_2)) \qquad \forall t \geq 0.$$

Then the solution $\bar{\psi}(\cdot; x_2) \in S(x_2)$ satisfies

$$
\frac{1}{C(x_2)} \leq \frac{1}{J(x_2, \bar{\psi})} = \int_0^{t_R(x_1)} \frac{1}{g(\bar{\psi}(t; x_2))} dt + \int_{t_R(x_1)}^\infty \frac{1}{g(\bar{\psi}(t; x_2))} dt
$$

$$
= \int_0^{t_R(x_1)} \frac{1}{g(\bar{\psi}(t; x_2))} dt + \int_0^\infty \frac{1}{g(\bar{\varphi}(t, \bar{\psi}(t_R(x_1), x_2)))} dt.
$$

The first term on the right-hand side can be estimated through

$$
\int_0^{t_R(x_1)} \frac{1}{g(\bar{\psi}(t; x_2))} dt \leq \int_0^{t_R(x_1)} \frac{1}{g(\bar{\psi}(t; x_2))} dt + \frac{\varepsilon}{2} \leq \frac{1}{J(x_1, \bar{\phi})} + \frac{\varepsilon}{2}
$$

$$
\leq \frac{1}{C(x_1)} + \frac{\varepsilon}{2} + \varepsilon_2
$$

and for the second term the inequalities

$$
\int_0^\infty \frac{1}{g(\bar{\varphi}(t, \bar{\psi}(t_R(x_1), x_2)))} dt \leq \frac{1}{C(\bar{\psi}(t_R(x_1), x_2))} + \varepsilon_2 \leq \frac{\varepsilon}{2} + \varepsilon_2
$$

hold. Here, the first inequality follows from the selection of $\bar{\varphi}$ and the second inequality follows from the definition of R after (6.94).

Combining both estimates, the inequality

$$
\frac{1}{C(x_2)} \leq \frac{1}{C(x_1)} + \varepsilon + 2\varepsilon_2.
$$

is obtained and since $\varepsilon_2 > 0$ was arbitrary, $C(x_2)^{-1} \leq C(x_1)^{-1} + \varepsilon$ is satisfied. Since x_1 and x_2 were arbitrary in $B_{\delta_\phi}(x)$, the same inequality holds if we exchange x_1 and x_2, i.e.,

$$
\left| \frac{1}{C(x_1)} - \frac{1}{C(x_2)} \right| \leq \varepsilon \qquad \forall\, x_1, x_2 \in B_{\delta_\phi}(x).
$$

This shows the required continuity estimate (6.94) for $\delta = \delta_\phi$. As discussed at the beginning of the proof, from the continuity of $C(\cdot)^{-1}$ for $x \neq 0$, continuity of C follows, which completes the proof. $\qquad\square$

The statements shown in the claims complete the proof of Proposition 6.16. $\quad\square$

Chapter 7
Auxiliary Results

Abstract In this chapter we collect auxiliary results used in the main proofs of the manuscript. The chapter is split in two sections. In the first part, results on comparison functions are derived. The results provide lower bounds used in the proofs in this monograph as counterparts to upper bounds contained in the literature. The second part collects known results used in Chap. 6.

Keywords Lyapunov methods · Differential inclusions · Stability of nonlinear systems · Instability of nonlinear systems · Stabilization/destabilization of nonlinear systems · Stabilizability and destabilizability

7.1 Comparison Results

As a first result we adapt the comparison principle in [67, Lemma A.4] or [36, Lemma 20], for example, to obtain a lower bound on trajectories in terms of $\mathcal{K}_\infty \mathcal{K}_\infty$-functions.

Lemma 7.1 *For any function $\rho \in \mathcal{P}$ there exists a $\kappa \in \mathcal{K}_\infty \mathcal{K}_\infty$ such that if $y : [0, T] \to \mathbb{R}$, $(T \in \mathbb{R}_{>0} \cup \{\infty\})$ is a locally absolutely continuous function which satisfies the differential inequality*

$$\dot{y}(t) \geq \lambda \rho(y(t)) \tag{7.1}$$

for almost all $t \in [0, T]$, for some $\lambda > 0$ with $y(0) = y_0 \in \mathbb{R}_{\geq 0}$ then

$$y(t) \geq \kappa(y_0, \lambda t), \qquad \forall \, t \in [0, T]. \tag{7.2}$$

Lemma 7.1 adapts the statements [67, Lemma A.4] and [36, Lemma 20] which instead of a lower bound in terms of a $\mathcal{K}_\infty \mathcal{K}_\infty$-function guarantee an upper bound in terms of a $\mathcal{K}\mathcal{L}$-function. The proof of Lemma 7.1 follows the lines of the presentation in [36, Lemma 20].

© The Author(s), under exclusive license to Springer Nature Switzerland AG 2021
P. Braun et al., *(In-)Stability of Differential Inclusions*,
SpringerBriefs in Mathematics,
https://doi.org/10.1007/978-3-030-76317-6_7

Proof We first demonstrate the case $\lambda = 1$. Let the assumptions of the lemma be satisfied for $\lambda = 1$. We define the function $\hat{\rho}(s) = \min\{s, \rho(s)\}$. Observe that

$$\dot{y}(t) \geq \lambda\rho(y(t)) \geq \lambda\hat{\rho}(y(t))$$

holds for all $t \in [0, T]$. For $s \in (0, \infty)$ we define the function

$$\eta(s) = \int_1^s \frac{1}{\hat{\rho}(\tau)}\, d\tau.$$

We observe that $\eta(s)$ is continuously differentiable and strictly increasing for $s \in (0, \infty)$. Due to the condition $s \geq \hat{\rho}(s)$ for all $s \in (0, 1)$ it holds that

$$\int_s^1 \frac{1}{\tau}\, d\tau \leq \int_s^1 \frac{1}{\hat{\rho}(\tau)}\, d\tau = -\eta(s),$$

which implies $\lim_{s \searrow 0} \eta(s) = -\infty$. For $s \in [1, \infty)$ it holds that

$$\int_1^s \frac{1}{\tau}\, d\tau \leq \int_1^s \frac{1}{\hat{\rho}(\tau)}\, d\tau = \eta(s),$$

and thus $\lim_{s \to \infty} \eta(s) = \infty$.

Hence, $\eta : (0, \infty) \to (-\infty, \infty)$ and, since η is continuous and strictly increasing, $\eta^{-1} : (-\infty, \infty) \to (0, \infty)$ is also continuous and strictly increasing. We define the function

$$\kappa(s, t) = \begin{cases} 0, & s = 0, \\ \eta^{-1}(\eta(s) + t), & s > 0, \end{cases} \quad \forall\, s, t \in \mathbb{R}_{\geq 0}.$$

Since $\eta^{-1}(\cdot)$ is unbounded it holds that $\eta^{-1}(\eta(s) + \cdot) - \eta^{-1}(\eta(s)) \in \mathcal{K}_\infty$ for all $s > 0$.

To show that κ is continuous at the origin assume for the sake of a contradiction that there exist a sequence $(s_i, t_i)_{i \in \mathbb{N}} \subset B_1(0)$ with $\lim_{i \to \infty}(s_i, t_i) \to (0, 0)$ and $\varepsilon > 0$ such that

$$\eta^{-1}(\eta(s_i) + t_i) > \varepsilon \quad \forall\, i \in \mathbb{N}. \tag{7.3}$$

Assuming that (7.3) is satisfied for all $i \in \mathbb{N}$ is not restrictive since we can always restrict the attention to a subsequence of $(s_i, t_i)_{i \in \mathbb{N}}$. The condition (7.3) is equivalent to

$$\eta(s_i) > \eta(\varepsilon) - t_i \geq \eta(\varepsilon) - 1 \quad \forall\, i \in \mathbb{N},$$

and hence a contradiction is obtained since $\eta(\varepsilon) - 1 \in \mathbb{R}$ and $\eta(s_i) \to -\infty$ for $i \to \infty$. Thus, continuity of κ at $(s, t) = (0, 0)$ follows.

With a similar argument it can be shown that $\lim_{s \to 0} \eta^{-1}(\eta(s) + t) = 0$ is satisfied for all $t \in \mathbb{R}_{\geq 0}$ fixed, and it holds that

$$\eta^{-1}(\eta(s) + t) \geq \eta^{-1}(\eta(s)) = s, \qquad \forall\, s, t \in \mathbb{R}_{\geq 0}.$$

From these properties it follows that $\kappa(\cdot, t) \in \mathcal{K}_\infty$ for all $t \in \mathbb{R}_{\geq 0}$ and we can conclude that $\kappa \in \mathcal{K}_\infty \mathcal{K}_\infty$.

If $y_0 = 0$, any function $\kappa \in \mathcal{K}_\infty \mathcal{K}_\infty$ satisfies $y(t) \geq 0 = \kappa(0, \lambda t)$ for all $t \in [0, T]$. In the case $y_0 > 0$, condition (7.1) ensures that $y(t) \geq y_0$, and thus $\hat{\rho}(y(t)) \neq 0$, for all $t \in [0, T]$. This implies that for all $y_0 > 0$ inequality (7.1) can be rewritten as

$$\frac{\dot{y}(t)}{\hat{\rho}(y(t))} \geq 1,$$

and integration over both sides leads to

$$\int_{y_0}^{y(t)} \frac{1}{\hat{\rho}(r)}\, dr = \int_0^t \frac{\dot{y}(\tau)}{\hat{\rho}(y(\tau))} d\tau \geq t.$$

Using the definition of the function η we obtain $\eta(y(t)) - \eta(y_0) \geq t$ or, equivalently,

$$y(t) \geq \eta^{-1}(\eta(y_0) + t),$$

which shows the assertion for $\lambda = 1$.

By taking the time rescaling $\tau = \lambda t$, we see that $\dot{y}(t) \geq \lambda \hat{\rho}(y(t))$ for almost all $t \in [0, T]$ becomes

$$\tfrac{d}{d\tau} y\left(\tfrac{\tau}{\lambda}\right) \geq \hat{\rho}\left(y\left(\tfrac{\tau}{\lambda}\right)\right),$$

for almost all $\tau \in [0, \lambda T]$. Following the steps above, we obtain a function $\kappa \in \mathcal{K}_\infty \mathcal{K}_\infty$ such that $y(\tau/\lambda) \geq \kappa(y_0, \tau)$ for all $\tau \in [0, \lambda T]$ and hence $y(t) \geq \kappa(y_0, \lambda t)$, for all $t \in [0, T]$. $\qquad \square$

As a next result we construct a lower bound equivalent to Sontag's lemma on $\mathcal{KL}$-estimates, again by reversing the inequalities. The original result, Sontag's lemma, was introduced in [65, Proposition 7].

Lemma 7.2 *For each $\kappa \in \mathcal{K}_\infty \mathcal{K}_\infty$ and $\lambda > 0$, there exist $\alpha, \gamma \in \mathcal{K}_\infty$ such that*

$$\alpha(\kappa(r, t)) \geq e^{\lambda t} \gamma(r) \qquad \forall\, (r, t) \in \mathbb{R}^2_{\geq 0}. \tag{7.4}$$

In contrast, Sontag's lemma shows that for $\beta \in \mathcal{KL}$ and $\lambda > 0$, there exist $\alpha, \gamma \in \mathcal{K}_\infty$ such that

$$\alpha(\beta(r, t)) \leq e^{\lambda t} \gamma(r) \qquad \forall\, (r, t) \in \mathbb{R}^2_{\geq 0}.$$

To prove Lemma 7.2 we first show that the following intermediate result is true.

Lemma 7.3 *For all $g \in \mathbb{R}$ and all $\kappa \in \mathcal{K}_\infty \mathcal{K}_\infty$ there exists $\hat{\alpha} \in \mathcal{K}_\infty$ such that*

$$\kappa\left(\hat{\alpha}(\tfrac{1}{s}), \ln(s) - g\right) \overset{\substack{s \to \infty \\ s \geq e^g}}{\longrightarrow} \infty \qquad \forall\, g \in \mathbb{R}.$$

Proof of Lemma 7.3 For each $s \geq 1$ fixed, we define $r(s) : [1, \infty) \to \mathbb{R}_0$ through the condition

$$\kappa\left(s^{-1}, \tfrac{1}{2} r(s)\right) = s. \tag{7.5}$$

Since κ is continuous and monotone, r depends continuously on s and is monotonically increasing. In particular, there exists $\rho \in \mathcal{K}_\infty$ such that

$$\kappa\left(s^{-1}, \tfrac{1}{2} \rho(s)\right) \geq s \qquad \forall\, s \geq 1.$$

As a next step we choose $\sigma \in \mathcal{K}_\infty$ with $\ln(\sigma(s)) \geq \rho(s)$ for all $s \geq 1$. Then $\kappa(\tfrac{1}{s}, \tfrac{1}{2}\ln(\sigma(s))) \geq s$ implies

$$\kappa\left(\tfrac{1}{\sigma^{-1}(s)}, \tfrac{1}{2} \ln(s)\right) \geq \sigma^{-1}(s).$$

We define

$$\hat{\alpha}(p) = \begin{cases} \dfrac{1}{\sigma^{-1}(\frac{1}{p})}, & \text{for } p > 0, \\[2mm] 0, & \text{for } p = 0. \end{cases}$$

Since $\sigma \in \mathcal{K}_\infty$, $\hat{\alpha}$ is a $\mathcal{K}_\infty$-function and for $s \neq 0$ it holds that

$$\kappa\left(\hat{\alpha}(s^{-1}), \tfrac{1}{2} \ln(s)\right) = \kappa\left(\frac{1}{\sigma^{-1}(s)}, \tfrac{1}{2} \ln(s)\right) \geq \sigma^{-1}(s) \overset{s \to \infty}{\longrightarrow} \infty.$$

Note that for all $g \in \mathbb{R}$, there exists $s' \in \mathbb{R}_{>1}$ such that

$$\ln(s) - g \geq \tfrac{1}{2} \ln(s), \qquad \forall\, s \geq s'.$$

This in particular implies that

$$\kappa\left(\hat{\alpha}(s^{-1}), \ln(s) - g\right) \geq \kappa\left(\hat{\alpha}(s^{-1}), \tfrac{1}{2} \ln(s)\right)$$

for s sufficiently large and which completes the proof.   □

Proof of Lemma 7.2 We prove that for $\hat{\alpha} \in \mathcal{K}_\infty$ from Lemma 7.3 there is $\alpha \in \mathcal{K}_\infty$ with

$$\alpha(\kappa(\hat{\alpha}(r), t)) \geq e^{\lambda t} r \qquad \forall (r, t) \in \mathbb{R}^2_{\geq 0}, \tag{7.6}$$

from which inequality (7.4) can be deduced. We first prove (7.6) for $\lambda = 1$. For $s \geq 1$ we consider the coordinate transformation $s = e^t$, implying $t = \ln(s)$.

For $s_{max}, r_{max} \in \mathbb{R}_{\geq 1}$ fixed, we define $q_{max} = s_{max} r_{max}$ and we consider the function $h : [0, q_{max}] \to \mathbb{R}$ defined as

$$h(q) = \min\{\kappa(\hat{\alpha}(r), \ln(s)) : q = sr, \ s \in [1, s_{max}], \ r \in [0, r_{max}]\}.$$

Since $\kappa(\cdot, \cdot)$, $\ln(\cdot)$ and $\hat{\alpha}(\cdot)$ are continuous functions and $[1, s_{max}] \times [0, r_{max}]$ is a compact domain, the minimum is attained and the function h is well defined. By definition, the function h satisfies $h(0) = 0$ and $h(q) > 0$ for all $q \in (0, q_{max}]$.

As a next step we show that h is strictly monotonically increasing. Let $q' > q$, $q, q' \in [0, q_{max}]$ and $q' = r' \cdot s'$. Since $0 \leq q < q'$ it holds that $s' \geq 1$ and there exists an $r < r'$ such that $q = r \cdot s'$. Since $\kappa(\cdot, s') \in \mathcal{K}_\infty$ (and $\hat{\alpha} \in \mathcal{K}_\infty$), the chain of inequalities

$$h(q') = \kappa(\hat{\alpha}(r'), \ln(s')) > \kappa(\hat{\alpha}(r), \ln(s')) \geq h(rs') = h(q)$$

is satisfied and thus h is strictly monotonically increasing.

We continue by showing that for all $q \in \mathbb{R}$ there exist $s_{max}, r_{max} \in \mathbb{R}$ such that

$$h(q) = \min\{\kappa(\hat{\alpha}(r), \ln(s)) : q = sr, \ s \in [1, s_{max}], \ r \in [0, r_{max}]\} \tag{7.7}$$
$$= \inf\{\kappa(\hat{\alpha}(r), \ln(s)) : q = sr, \ s \geq 1, \ r \geq 0\}, \tag{7.8}$$

i.e., the restriction to a compact domain is not necessary.

To this end, for the sake of a contradiction, assume there exist $q, c \in \mathbb{R}_{\geq 0}$ and sequences $(s_i)_{i \in \mathbb{N}}$, $(r_i)_{i \in \mathbb{N}}$ with the properties $q = r_i s_i$ for all $i \in \mathbb{N}$ and $s_i \to \infty$ for $i \to \infty$ such that

$$\kappa(\hat{\alpha}(r_i), \ln(s_i)) \leq c \tag{7.9}$$

is satisfied for all $i \in \mathbb{N}$. The other case, i.e., $r_i \to \infty$ for $i \to \infty$ (and $(s_i)_{i \in \mathbb{N}} \subset \mathbb{R}_{\geq 1}$ arbitrary) leads trivially to a contradiction since $\kappa(\hat{\alpha}(\cdot), \ln(1))$ is a $\mathcal{K}_\infty$-function and $\kappa(\hat{\alpha}(r_i), \ln(s_i)) \geq \kappa(r_i, 0) \to \infty$.

For $q = r_i s_i$ with $s_i \to \infty$ for $i \to \infty$ we define $p_i = q/r_i$ which implies $s_i = qp_i$ and $p_i \to \infty$. Thus condition (7.9) can be rewritten as

$$\kappa\left(\hat{\alpha}(p_i^{-1}), \ln(qp_i)\right) = \kappa\left(\hat{\alpha}(p_i^{-1}), \ln(p_i) - g\right) \leq c \tag{7.10}$$

for all $i \in \mathbb{N}$ and where g is defined as $g = -\ln(q)$. This is however a contradiction to Lemma 7.3 and thus h in (7.8) is well defined and, in particular, $h(q) \to \infty$ for $q \to \infty$.

As a last step, we need to show that h is continuous. Let $q_1 > q_0 \geq 0$ and $q_1 = r_1 s_1$ and $q_0 = r_0 s_0$. We can write

$$q_1 = \frac{q_1}{s_0} s_0 \quad \text{and} \quad r_0 = \frac{q_0}{s_0}.$$

It holds that

$$0 \leq h(q_1) - h(q_0) \leq \kappa \left(\hat{\alpha} \left(\tfrac{q_1}{s_0} \right), \ln(s_0) \right) - \kappa (\hat{\alpha} \left(\tfrac{q_0}{s_0} \right), \ln(s_0)) \to 0$$

for $q_1 \searrow q_0$. Similarly, for $q_0 > q_1 \geq 0$ the estimate

$$0 \leq h(q_0) - h(q_1) \leq \kappa \left(\hat{\alpha} \left(\tfrac{q_0}{s_1} \right), \ln(s_1) \right) - \kappa \left(\hat{\alpha} \left(\tfrac{q_1}{s_1} \right), \ln(s_1) \right) \to 0$$

is satisfied for $q_1 \nearrow q_0$ and continuity of h follows. Moreover, $h \in \mathcal{K}_\infty$,

$$h(q) = h(rs) \leq \kappa(\hat{\alpha}(r), \ln(s))$$

for all $q \in \mathbb{R} \geq 0$, for all $r \geq 0$, $s \geq 1$ with $q = rs$ and thus

$$re^t \leq h^{-1}(\kappa(\hat{\alpha}(r), t)).$$

In particular, for $\lambda = 1$, inequality (7.6) follows with $\alpha = h^{-1}$. Inequality (7.6) for $\lambda > 0$ follows from the coordinate transformation $s = e^{\lambda t}$ instead of $s = e^t$.

Finally, inequality (7.4) is obtained by defining $\gamma = \hat{\alpha}^{-1}$, since applying (7.6) with $\gamma(r)$ in place of r yields

$$\alpha(\kappa(r, t)) = \alpha(\kappa(\hat{\alpha}(\gamma(r)), t)) \leq e^{\lambda t} \gamma(r).$$

$\square$

The next lemma is a slight variation of [69, Lemma 16] by changing some inequalities in the original result. The proof follows by making minimal changes to the proof of [69, Lemma 16].

Lemma 7.4 *Let $O \subset \mathbb{R}^n$ be open and let the three functions $\alpha : O \to \mathbb{R}$ and $\mu, \nu : O \to \mathbb{R}_{>0}$ be continuous. Suppose $C : O \to \mathbb{R}$ is locally Lipschitz on O, and (2.1) satisfies Assumption 2.1 on O and is locally Lipschitz on O, and for almost all $x \in O$,*

$$\min_{w \in F(x)} \langle \nabla C(x), w \rangle \geq \alpha(x).$$

Then there exists a smooth function $C_s : O \to \mathbb{R}$ such that, for all $x \in O$,

$$|C(x) - C_s(x)| \leq \mu(x) \tag{7.11}$$

and

$$\min_{w \in F(x)} \langle \nabla C_s(x), w \rangle \geq \alpha(x) - \nu(x). \tag{7.12}$$

The proof of Lemma 7.4 relies on Claim 7.5 whose proof is given first. Additionally, the following notation is needed, assuming that O and α, μ, σ are defined according to Lemma 7.4. With $\psi : \mathbb{R}^n \to [0, \infty)$ we denote a smooth function, which vanishes outside of the unit disk and satisfies

$$\int_{\mathbb{R}^n} \psi(\xi) d\xi = 1.$$

Moreover, for $\sigma \in (0, 1]$ and for those x such that $\overline{B}_\sigma(x) \subset O$, we define the functions

$$C_\sigma(x) = \int_{\mathbb{R}^n} C(x + \sigma\xi)\psi(\xi)d\xi, \tag{7.13}$$

$$\alpha_\sigma(x) = \int_{\mathbb{R}^n} \alpha(x + \sigma\xi)\psi(\xi)d\xi. \tag{7.14}$$

Claim 7.5 *For each compact set $\mathcal{A} \subset O$ and $\varepsilon > 0$, there exists $\sigma_0 > 0$ such that, for all $\sigma \in (0, \sigma_0)$, the functions C_σ and α_σ are smooth on $\mathcal{A}$ and for all $x \in \mathcal{A}$, we have*

$$\max_{w \in F(x)} \langle \nabla C_\sigma(x), w \rangle \geq \alpha(x) - \varepsilon, \tag{7.15}$$

$$|C(x) - C_\sigma(x)| \leq \varepsilon, \tag{7.16}$$

$$|\alpha(x) - \alpha_\sigma(x)| \leq \tfrac{1}{2}\varepsilon. \tag{7.17}$$

Claim 7.5 (and its proof) is a slight variation of [69, Claim 6], where the sign in front of (7.15) is changed. For completeness, we report this modified proof here.

Proof of Claim 7.5 The functions C_σ and α_σ in (7.13) and (7.14), respectively, are defined in the same way as in the proof [69, Lemma 16]. Thus the smoothness of C_σ and α_σ as well as (7.16), (7.17) follow from the proof [69, Lemma 16] and where the authors refer to [72, Proposition I.8] for a detailed reference.

To obtain (7.15), we follow [69, Lemma 16], which, in turn, refers to [23, Lemma 5.1] and [47, Lemma B.5]. Let $\rho > 0$ be such that $\mathcal{A} + B_\rho(0) \subset O$ and let L be a Lipschitz constant for F on $\mathcal{A} + \overline{B}_\rho(0)$. Let $M_1 > 0$ satisfy $|\xi| \leq M_1$ for all $\xi \in \mathcal{A}$ and let $M_2 > 0$ satisfy $|\nabla C(\xi)| \leq M_2$ for almost all $\xi \in \mathcal{A} + \overline{B}_\rho(0)$. Define

$$\sigma_0 = \min\left\{\rho, \frac{\varepsilon}{2LM_1M_2}\right\}$$

and consider $\sigma \in (0, \sigma_0)$. Let $x \in \mathcal{A}$ and $\eta \in F(x)$. Given $\xi \in \overline{B}_1(0)$, let $g_\sigma(\xi)$ be the closest point in $F(x + \sigma\xi)$ to η. Then $g_\sigma : \overline{B}_1(0) \to \mathbb{R}^n$ is continuous and

$$g_\sigma(\xi) \in F(x + \sigma\xi), \qquad |g_\sigma(\xi) - \eta| \le L\sigma|\xi| \quad \forall \xi \in \overline{B}_1(0).$$

Using the Lebesgue dominated convergence theorem, we get

$$\langle \nabla C_\sigma(x), \eta \rangle = \int_{\mathbb{R}^n} \langle \nabla C(x + \sigma\xi), \eta \rangle \psi(\xi) d\xi.$$

Then we have

$$
\begin{aligned}
\langle \nabla C_\sigma, \eta \rangle &= \int_{\mathbb{R}^n} \langle \nabla C(x + \sigma\xi), \eta \rangle \psi(\xi) d\xi \\
&= \int_{\mathbb{R}^n} \langle \nabla C(x + \sigma\xi), g_\sigma(\xi) \rangle \psi(\xi) d\xi + \int_{\mathbb{R}^n} \langle \nabla C(x + \sigma\xi), \eta - g_\sigma(\xi) \rangle \psi(\xi) d\xi \\
&\ge \int_{\mathbb{R}^n} \alpha(x + \sigma\xi)\psi(\xi) d\xi - L\sigma \int_{\mathbb{R}^n} |\nabla C(x + \sigma\xi)||\xi|\psi(\xi) d\xi \\
&= \alpha_\sigma(x) - L\sigma \int_{\mathbb{R}^n} |\nabla C(x + \sigma\xi)||\xi|\psi(\xi) d\xi \\
&\ge \alpha(x) - \frac{\varepsilon}{2} - L\sigma M_1 M_2 \int_{\mathbb{R}^n} \psi(\xi) d\xi \ge \alpha(x) - \varepsilon.
\end{aligned}
$$

which completes the proof. $\square$

Proof of Lemma 7.4 Let $\{\mathcal{U}_i\}_{i=1}^\infty$ be a locally finite open cover of O with $\overline{\mathcal{U}}_i \subset O$ being compact for all $i \in \mathbb{N}$. Additionally, let $\kappa_i, i \in \mathbb{N}$, be a smooth partition of unity on O subordinate to $\{\mathcal{U}_i\}_{i=1}^\infty$. This means, for each $x \in O$ there is a neighborhood of x that intersects only with a finite number of sets of $\mathcal{U}_i$ and the functions $\kappa_i : O \to [0, 1]$ satisfy

$$\kappa_i(x) > 0 \quad \Longrightarrow \quad x \in \mathcal{U}_i,$$

$$\sum_{i=1}^\infty \kappa_i(x) = 1 \quad \forall x \in O. \tag{7.18}$$

Condition (7.18) implies that

$$\sum_{i=1}^\infty \langle \nabla \kappa_i(x), \eta \rangle = 0, \qquad \forall\, (x, \eta) \in O \times \mathbb{R}^n \tag{7.19}$$

(since $\sum_{i=1}^\infty \kappa_i$ is a constant function). We define $\varepsilon_i, q_i, i \in \mathbb{N}$ through the conditions

$$\varepsilon_i = \inf_{\xi \in \mathcal{U}_i} \min\{\mu(\xi), \nu(\xi)\}, \quad \text{and} \quad q_i = \max_{\xi \in \overline{\mathcal{U}}_i,\, \eta \in F(\xi)} |\nabla \kappa_i(\xi)||\eta|.$$

From Claim 7.5 we know that for each $i \in \mathbb{N}$, there exists σ_i such that C_{σ_i} is smooth on $\overline{\mathcal{U}}_i$ and such that the conditions

$$|C(x) - C_{\sigma_i}(x)| \leq \frac{\varepsilon_i}{2^{(i+1)}(1+q_i)} \qquad\qquad \forall x \in \overline{\mathcal{U}}_i,$$

$$\langle \nabla C_{\sigma_i}(x), \eta \rangle \geq \alpha(x) - \frac{\varepsilon_i}{2^{(i+1)}(1+q_i)} \geq \alpha(x) - \frac{\varepsilon_i}{2} \qquad \forall \, \eta \in F(x)$$

are satisfied.

Define $C_s(x) = \sum_{i=1}^{\infty} \kappa_i(x) C_{\sigma_i}(x)$. Then, C_s is smooth on O. Additionally, for all $x \in O$ we define the index set $\mathcal{I}_x = \{j \in \mathbb{N} : x \in \mathcal{U}_j\}$. Then, it holds that

$$|C(x) - C_s(x)| \leq \sum_{i=1}^{\infty} \kappa_i(x)|C(x) - C_{\sigma_i}(x)| \leq \max_{j \in \mathcal{I}_x} \varepsilon_j \leq \mu(x).$$

Finally, with (7.18) and (7.19), we get, for all $x \in O$,

$$
\begin{aligned}
\langle \nabla C_s(x), \eta \rangle &= \sum_{i=1}^{\infty} \langle \nabla \kappa_i(x), \eta \rangle C_{\sigma_i}(x) + \sum_{i=1}^{\infty} \kappa_i(x) \langle \nabla C_{\sigma_i}(x), \eta \rangle \\
&= \sum_{i=1}^{\infty} \langle \nabla \kappa_i(x), \eta \rangle (C_{\sigma_i}(x) - C(x)) + \sum_{i=1}^{\infty} \kappa_i(x) \langle \nabla C_{\sigma_i}(x), \eta \rangle \\
&\geq \sum_{i \in \mathcal{I}_x} \left[-q_i \frac{\varepsilon_i}{2^{(2+1)}(1+q_i)} + \kappa_i(x) \left(\alpha(x) - \frac{\varepsilon_i}{2} \right) \right] \\
&\geq \alpha(x) - \frac{1}{2} \sum_{i \in \mathcal{I}_x} \kappa_i(x) \max_{j \in \mathcal{I}_x} \varepsilon_j - \frac{1}{2} \max_{j \in \mathcal{I}_x} \varepsilon_j \sum_{i \in \mathcal{I}_x} \left(\frac{q_i}{1+q_i} \frac{1}{2^i} \right) \\
&\geq \alpha(x) - \max_{j \in \mathcal{I}_x} \varepsilon_j \geq \alpha(x) - \nu(x)
\end{aligned}
$$

Since $\eta \in F(x)$ was arbitrary, the result follows. $\qquad\qquad\qquad\square$

7.2 Known Results Used in Chap. 6

This section collects several results from the literature that were used in Chap. 6.

Lemma 7.6 ([36, Lemma 24], [39, Lemma 18]) *For each $\rho \in \mathcal{P}$, there exists $\alpha \in \mathcal{K}_\infty$ which is locally Lipschitz on $\mathbb{R}_{\geq 0}$ and continuously differentiable on $\mathbb{R}_{>0}$ and satisfies*

$$\alpha(s) \leq \rho(s)\alpha'(s), \qquad \forall \, s \in \mathbb{R}_{\geq 0}.$$

Lemma 7.6 is used in Sect. 3.3 to manipulate the right-hand side of the decrease condition (increase condition) of a Lyapunov function (Chetaev function).

Lemma 7.7 ([30, §7, Theorem 3], [69, Lemma 4]) *Consider the differential inclusion (2.1) satisfying Assumption 2.1 and suppose that $\mathcal{A} \subset \mathbb{R}^n$, compact, and $T > 0$ are such that all solutions $\phi \in \mathcal{S}(x)$, $x \in \mathcal{A}$ satisfy $|\phi(t; x)| < \infty$ for all $t \in [0, T]$. Then the set*

$$\mathcal{R}_{\leq T}(\mathcal{A}) = \{\xi \in \mathbb{R}^n | \xi = \phi(t; x), t \in [0, T], x \in \mathcal{A}, \phi \in \mathcal{S}(x)\}$$

is a compact subset of $\mathbb{R}^n$ and the set

$$S[0, T](\mathcal{A}) = \{\phi : [0, T] \to \mathbb{R}^n | \phi \in \mathcal{S}(x), \ x \in \mathcal{A}\}$$

is a compact set in the metric of uniform convergence.

Here, $S[0, T](\mathcal{A})$ denotes the set of solutions $\mathcal{S}(\mathcal{A})$ restricted to the time interval $[0, T]$. Lemma 7.7 is used in the proofs of Theorems 3.13, 3.15 and 4.11 to be able to restrict attention to a compact set and, for any sequence in $\mathcal{A}$ or $\mathcal{R}_{\leq T}(\mathcal{A})$, to conclude the existence of a converging subsequence.

Lemma 7.8 ([69, Lemma 6]) *Consider the differential inclusion (2.1). Suppose (2.1) is forward complete on $\mathbb{R}^n$ and F satisfies Assumption 2.1. For each triple $(T, \varepsilon, \mathcal{A})$ where $T > 0$, $\varepsilon > 0$ and $\mathcal{A} \subset \mathbb{R}^n$ compact, there exists $\delta > 0$ such that for every maximal solution $\phi_\delta(\cdot; x_\delta)$ of*

$$\dot{x} \in F_\delta(x) = \overline{\mathrm{conv}}(F(x + \overline{B}_\delta)) + \overline{B}_\delta$$

with $x_\delta \in \mathcal{A} + \overline{B}_\delta$, there exists a solution $\phi(\cdot; x) \in \mathcal{S}(x)$ of (2.1) with $x \in \mathcal{A}$ and $|x - x_\delta| \leq \varepsilon$ such that, for all $t \in [0, T]$,

$$\left| |\phi_\delta(t; x_\delta)| - |\phi(t; x)| \right| \leq \varepsilon.$$

Lemma 7.9 ([69, Lemma 8]) *Consider the differential inclusion (2.1) satisfying Assumption 2.1. Let $\Delta : \mathbb{R}^n \to \mathbb{R}_{\geq 0}$ be a positive function. Then there exists a set valued-map F_L satisfying Assumptions 2.1, 2.2 and*

$$F(x) \subset F_L(x) \subset F_{\Delta(x)}(x) = \overline{\mathrm{conv}}(F(x + \overline{B}_{\Delta(x)}) + \overline{B}_{\Delta(x)}.$$

Lemmas 7.8 and 7.9 show that for every differential inclusion satisfying Assumption 2.1 there exists a differential inclusion arbitrarily close which satisfies Assumptions 2.1 and 2.2. The statements are used in the proof of Theorem 3.15 to be able to use a differential inclusion with a Lipschitz continuous right-hand side.

Lemma 7.10 ([69, Lemma 10]) *Consider the differential inclusion (2.1). Let F satisfy the Assumption 2.1 on $\mathbb{R}^n \backslash \{0\}$ and be locally Lipschitz on the open set $O \subset \mathbb{R}^n \backslash \{0\}$. For each $T > 0$ and each compact set $\mathcal{A} \subset O$, there exist L and $\delta > 0$ such that, for each $x \in \mathcal{A}$, each $\phi \in \mathcal{S}(x)$ and each ξ satisfying $|x - \xi| \leq \delta$, there exists $\psi \in \mathcal{S}(\xi)$ with the property*

$$|\phi(t; x) - \psi(t; \xi)| \le L|x - \xi| \quad \forall t \in [0, T_x]$$

where $T_x \in [0, T]$ is such that $\phi(t; x) \in \mathcal{A}$ for all $t \in [0, T_x]$.

Lemma 7.10 is used to show lower semicontinuity and Lipschitz continuity of the Chetaev function in the proof of Theorem 3.15. For $\varepsilon > 0$ and L and δ defined such that $\varepsilon \ge L\delta$, Lemma 7.10 immediately implies the following minimal variation of this result which is used in the proof of Theorem 4.11 to conclude continuity of the candidate control Chetaev function.

Corollary 7.11 *Consider the differential inclusion (2.1). Let F satisfy the Assumption 2.1 on $\mathbb{R}^n\backslash\{0\}$ and be locally Lipschitz on the open set $O \subset \mathbb{R}^n\backslash\{0\}$. For each $T > 0$, each compact set $\mathcal{A} \subset O$ and each $\varepsilon > 0$, there exists $\delta > 0$ such that, for each $x \in \mathcal{A}$, each $\phi \in \mathcal{S}(x)$ and each ξ satisfying $|x - \xi| \le \delta$, there exists $\psi \in \mathcal{S}(\xi)$ with the property*

$$|\phi(t; x) - \psi(t; \xi)| \le \varepsilon \quad \forall t \in [0, T_x]$$

where $T_x \in [0, T]$ is such that $\phi(t; x) \in \mathcal{A}$ for all $t \in [0, T_x]$.

Lemma 7.12 ([69, Lemma 11]) *Consider the differential inclusion (2.1) satisfying Assumptions 2.1 and 2.2 and let $x \in \mathbb{R}^n\backslash\{0\}$. Then for each $v \in F(x)$ there exists a solution $\phi \in \mathcal{S}(x)$ satisfying*

$$\phi(t; x) = x + t(v + r(t)), \quad \forall t \in [0, T)$$

for some $T > 0$ and for some function $r(\cdot)$ that is continuous on $[0, T)$ and satisfies $\lim_{t \searrow 0} r(t) = 0$.

The result is used in the proof of Theorem 4.11 to construct a direction which satisfies the increase condition (4.8).

Lemma 7.13 ([69, Lemma 14], [25, Corollary 3.7]) *Let $V : O \to \mathbb{R} \cup \{\infty\}$ ($O \subset \mathbb{R}^n$) be lower semicontinuous. Let $\mathcal{U} \subset O$ be open and convex. Then, the function V is Lipschitz with Lipschitz constant M on $\mathcal{U}$ if and only if*

$$D_+ V(x; v) \le M|v| \quad \forall x \in \mathcal{U}, \quad \forall v \in \mathbb{R}^n.$$

Lemma 7.13 is used in the proof of Theorem 3.15 to conclude Lipschitz continuity of the candidate Chetaev function. Similarly, the next result is used to construct a smooth Chetaev function based on a Lipschitz continuous Chetaev function in the proof of Theorem 3.15.

Lemma 7.14 ([69, Lemma 17]) *Assume that $V : \mathbb{R}^n \to \mathbb{R}_{\ge 0}$ is continuous, $V : \mathbb{R}^n\backslash\{0\} \to \mathbb{R}_{>0}$ is smooth and $V(0) = 0$. Then there exists a function $\rho \in \mathcal{K}_\infty$, smooth on $(0, \infty)$ and $\rho' \in \mathcal{K}_\infty$ and with the property $\rho(s) \le s\rho'(s)$ for all $s \ge 0$, such that $V_s = \rho \circ V$ is smooth on $\mathbb{R}^n$.*

References

1. Ames, A.D., Coogan, S., Egerstedt, M., Notomista, G., Sreenath, K., Tabuada, P.: Control barrier functions: theory and applications. In: 18th European Control Conference, pp. 3420–3431 (2019)
2. Ames, A.D., Xu, X., Grizzle, J.W., Tabuada, P.: Control barrier function based quadratic programs for safety critical systems. IEEE Trans. Autom. Control $62(8)$, 3861–3876 (2017)
3. Andriano, V., Bacciotti, A., Beccari, G.: Global stability and external stability of dynamical systems. Nonlinear Anal: Theory, Methods Appl. $28(7)$, 1167–1185 (1997)
4. Angeli, D., Efimov, D.: Characterizations of input-to-state stability for systems with multiple invariant sets. IEEE Trans. Autom. Control $60(12)$, 3242–3256 (2015)
5. Artstein, Z.: Stabilization with relaxed controls. Nonlinear Anal. $7(11)$, 1163–1173 (1983)
6. Aubin, J.-P., Frankowska, H.: Set-Valued Analysis. Springer Science & Business Media, Berlin (2009)
7. Bacciotti, A., Rosier, L.: Liapunov Functions and Stability in Control Theory. Springer Science & Business Media, Berlin (2006)
8. Barbashin, E.A., Krasovskii, N.N.: On the stability of motion in the large. Dokl. Acad. Nauk. 86, 453–456 (1952)
9. Bhatia, N.P., Szegö, G.P.: Stability Theory of Dynamical Systems. Springer Science & Business Media, Berlin (1967)
10. Braun, P., Grüne, L., Kellett, C.M.: Feedback design using nonsmooth control Lyapunov functions: A numerical case study for the nonholonomic integrator. In: Proceedings of the 56th IEEE Conference on Decision and Control, pp. 4890–4895 (2017)
11. Braun, P., Grüne, L., Kellett, C.M.: Complete instability of differential inclusions using Lyapunov methods. In: Proceedings of the 57th IEEE Conference on Decision and Control, pp. 718–724 (2018)
12. Braun, P., Kellett, C.M.: Comment on "Stabilization with guaranteed safety using control Lyapunov-barrier function." Automatica 122 (2020)
13. Braun, P., Kellett, C.M., Zaccarian, L.: Complete control Lyapunov functions: stability under state constraints. IFAC-PapersOnLine $52(16)$, 358–363 (2019). 11th IFAC Symposium on Nonlinear Control Systems
14. Braun, P., Kellett, C.M., Zaccarian, L.: Uniting control laws: On obstacle avoidance and global stabilization of underactuated linear systems. In: Proceedings of the IEEE 58th Conference on Decision and Control, pp. 8154–8159 (2019)

© The Author(s), under exclusive license to Springer Nature Switzerland AG 2021
P. Braun et al., *(In-)Stability of Differential Inclusions*,
SpringerBriefs in Mathematics,
https://doi.org/10.1007/978-3-030-76317-6

15. Braun, P., Kellett, C.M., Zaccarian, L.: Explicit construction of stabilizing robust avoidance controllers for linear systems with drift. IEEE Trans. Autom. Control **66**(2), 595–610 (2021)
16. Brockett, R.W.: Asymptotic stability and feedback stabilization. In: Brockett, R.W., Millman, R.S., Sussman, H.J. (eds.) Differential Geometric Control Theory, pp. 181–191. Birkhauser, Boston (1982)
17. Camilli, F., Grüne, L., Wirth, F.: Control Lyapunov functions and Zubov's method. SIAM J. Control. Optim. **47**(1), 301–326 (2008)
18. Cannarsa, P., Sinestrari, C.: Semiconcave Functions, Hamilton-Jacobi Equations, and Optimal Control, vol. 58. Springer Science & Business Media, Berlin (2004)
19. Chetaev, N.G.: The Stability of Motion. Pergamon Press, Oxford (1962). Translated from the 2nd Edition in Russian of 1956
20. Clarke, F.: Lyapunov functions and discontinuous stabilizing feedback. Annu. Rev. Control. **35**(1), 13–33 (2011)
21. Clarke, F.H., Ledyaev, Y.S., Rifford, L., Stern, R.J.: Feedback stabilization and Lyapunov functions. SIAM J. Control. Optim. **39**(1), 25–48 (2000)
22. Clarke, F.H., Ledyaev, Y.S., Sontag, E.D., Subbotin, A.I.: Asymptotic controllability implies feedback stabilization. IEEE Trans. Autom. Control **42**(10), 1394–1407 (1997)
23. Clarke, F.H., Ledyaev, Y.S., Stern, R.J.: Asymptotic stability and smooth Lyapunov functions. J. Differential Equ. **149**(1), 69–114 (1998)
24. Clarke, F.H., Ledyaev, Y.S., Stern, R.J., Wolenski, P.R.: Nonsmooth Analysis and Control Theory, vol. 178. Springer, Berlin (1998)
25. Clarke, F.H., Stern, R.J., Wolenski, P.R.: Subgradient criteria for monotonicity, the Lipschitz condition, and convexity. Can. J. Math. **45**(6), 1167–1183 (1993)
26. Efimov, D., Perruquetti, W.: Oscillating system design applying universal formula for control. In: Proceedings of the 50th IEEE Conference on Decision and Control and European Control Conference, pp. 1747–1752 (2011)
27. Efimov, D., Perruquetti, W., Petreczky, M.: On necessary conditions of instability and design of destabilizing controls. In: Proceedings of the 53rd IEEE Conference on Decision and Control, pp. 3915–3917 (2014)
28. Efimov, D.V., Fradkov, A.L.: Oscillatority of nonlinear systems with static feedback. SIAM J. Control. Optim. **48**(2), 618–640 (2009)
29. Evans, L.C., Gariepy, R.F.: Measure Theory and Fine Properties of Functions. CRC Press, New York (2015)
30. Filippov, A.F.: Differential Equations with Discontinuous Righthand Sides. Kluwer Academic Publishers, Amsterdam (1988)
31. Forni, P., Angeli, D.: Smooth Lyapunov functions for multistable differential inclusions. IFAC-PapersOnLine **50**(1), 1661–1666 (2017). 20th IFAC World Congress
32. Goebel, R., Sanfelice, R.G., Teel, A.R.: Hybrid Dynamical Systems: Modeling, Stability, and Robustness. Princeton University Press, Princeton (2012)
33. Guerra, M., Efimov, D., Zheng, G., Perruquetti, W.: Avoiding local minima in the potential field method using input-to-state stability. Control. Eng. Pract. **55**, 174–184 (2016)
34. Hahn, W.: Theory and Application of Liapunov's Direct Method. Prentice-Hall, New Jersey (1963)
35. Hahn, W.: Stability of Motion, vol. 138. Springer, Berlin (1967)
36. Kellett, C.M.: A compendium of comparision function results. Math. Controls, Signals Syst. **26**(3), 339–374 (2014)
37. Kellett, C.M.: Classical converse theorems in Lyapunov's second method. Discret. Contin. Dyn. Syst. - B **20**(8), 2333–2360 (2015)
38. Kellett, C.M., Teel, A.R.: Uniform asymptotic controllability to a set implies locally Lipschitz control-Lyapunov function. In: Proceedings of the 39th IEEE Conference on Decision and Control, pp. 3994–3999 (2000)
39. Kellett, C.M., Teel, A.R.: Weak converse Lyapunov theorems and control Lyapunov functions. SIAM J. Control. Optim. **42**(6), 1934–1959 (2004)
40. Khalil, H.K.: Nonlinear Systems, 3rd edn. Prentice Hall, New Jersey (1996)

41. Khatib, O.: Real-time obstacle avoidance for manipulators and mobile robots. In: Proceedings of the IEEE International Conference on Robotics and Automation, pp. 500–505 (1985)
42. Khatib, O.: Real-Time Obstacle Avoidance for Manipulators and Mobile Robots, pp. 396–404. Springer New York (1990)
43. Krasovskii, N.N.: Stability of Motion: Applications of Lyapunov's Second Method to Differential Systems and Equations with Delay. Stanford University Press, Stanford (1963)
44. Krasowski, N.N.: The converse of the theorem of K. P. Persidskij on uniform stability. Prikladnaja Matematika I Mehanica **19**, 273–278 (1955). (in Russian)
45. Lakshmikantham, V.: On Massera type converse theorem in terms of two different measures. Technical report, University of Texas at Arlington (1975)
46. Lakshmikantham, V., Leela, S., Martynyuk, A.A.: Stability Analysis of Nonlinear Systems. Springer, Berlin (1989)
47. Lin, Y., Sontag, E.D., Wang, Y.: A smooth converse Lyapunov theorem for robust stability. SIAM J. Control. Optim. **34**(1), 124–160 (1996)
48. Loría, A., Panteley, E.: Stability, Told by Its Developers, pp. 199–258. Springer, Berlin (2006)
49. Loría, A., Panteley, E.: Stability, as told by its developers. IFAC-PapersOnLine **50**(1), 5219–5230 (2017). 20th IFAC World Congress
50. Lyapunov, A.M.: The general problem of the stability of motion. Math. Soc. of Kharkov (1892). (Russian). (English Translation, Int. J. Control **55**, 531–773 (1992))
51. Malisoff, M., Mazenc, F.: Constructions of Strict Lyapunov Functions. Springer Science & Business Media, Berlin (2009)
52. Malkin, I.G.: On the question of the reciprocal of Lyapunov's theorem on asymptotic stability. (Russian) Prikladnaya Matematika i Mekhanika **18**, 129–138 (1954)
53. Massera, J.L.: Contributions to stability theory. Ann. Math. 182–206 (1956)
54. Michel, A.N., Hou, L., Liu, D.: Stability of Dynamical Systems. Springer, Berlin (2008)
55. Movchan, A.A.: Stability of processes with respect to two metrics. J. Appl. Math. Mech. **24**(6), 1506–1524 (1960)
56. Orlov, Y.: Nonsmooth Lyapunov Analysis in Finite and Infinite Dimensions. Springer, Berlin (2020)
57. Paternain, S., Koditschek, D.E., Ribeiro, A.: Navigation functions for convex potentials in a space with convex obstacles. IEEE Trans. Autom. Control **63**(9), 2944–2959 (2018)
58. Polyakov, A., Fridman, L.: Stability notions and Lyapunov functions for sliding mode control systems. J. Franklin Inst. **351**(4), 1831–1865 (2014). Special Issue on 2010-2012 Advances in Variable Structure Systems and Sliding Mode Algorithms
59. Rifford, L.: Existence of Lipschitz and semiconcave control-Lyapunov functions. SIAM J. Control. Optim. **39**(4), 1043–1064 (2000)
60. Rouche, N., Habets, P., Laloy, M.: Stability Theory by Liapunov's Direct Method, vol. 4. Springer, Berlin (1977)
61. Seah, S.W.: Existence of solutions and asymptotic equilibrium of multivalued differential systems. J. Math. Anal. Appl. **89**(2), 648–663 (1982)
62. Serrin, J., Varberg, D.E.: A general chain rule for derivatives and the change of variables formula for the Lebesgue integral. Am. Math. Mon. **76**(5), 514–520 (1969)
63. Sontag, E.D.: A Lyapunov-like characterization of asymptotic controllability. SIAM J. Control. Optim. **21**, 462–471 (1983)
64. Sontag, E.D.: A 'universal' construction of Artstein's theorem on nonlinear stabilization. Syst. Control Lett. **13**(2), 117–123 (1989)
65. Sontag, E.D.: Comments on integral variants of ISS. Syst. Control Lett. **34**(1–2), 93–100 (1998)
66. Sontag, E.D., Sussman, H.J.: Nonsmooth control-Lyapunov functions. In: Proceedings of the 34th Conference on Decision and Control, pp. 2799–2805 (1995)
67. Sontag, E.D., Wang, Y.: Lyapunov characterizations of input to output stability. SIAM J. Control. Optim. **39**(1), 226–249 (2000)
68. Taniguchi, T.: Global existence of solutions of differential inclusions. J. Math. Anal. Appl. **166**(1), 41–51 (1992)

69. Teel, A.R., Praly, L.: A smooth Lyapunov function from a class-$\mathcal{KL}$ estimate involving two positive semidefinite functions. ESAIM: Control, Optim. Calculus Vari. **5**, 313–367 (2000)
70. Vinograd, R.È.: The inadequacy of the method of characteristic exponents when applied to non-linear equations. In: Doklady Akademii Nauk, vol. 114, pp. 239–240. Russian Academy of Sciences (1957)
71. Yan, M.: Introduction to Topology. Walter de Gruyter GmbH & Co KG (2016)
72. Yosida, K.: Functional Analysis. Springer, Berlin (1986)